Synthesis Lectures on Mechanical Engineering

This series publishes short books in mechanical engineering (ME), the engineering branch that combines engineering, physics and mathematics principles with materials science to design, analyze, manufacture, and maintain mechanical systems. It involves the production and usage of heat and mechanical power for the design, production and operation of machines and tools. This series publishes within all areas of ME and follows the ASME technical division categories.

Luis Gonzalo Mejía Cañas

Introduction to the Theory of Vehicular Collisions

 Springer

Luis Gonzalo Mejía Cañas
Independent
Medellín, Colombia

ISSN 2573-3168   ISSN 2573-3176 (electronic)
Synthesis Lectures on Mechanical Engineering
ISBN 978-3-031-62354-7   ISBN 978-3-031-62355-4 (eBook)
https://doi.org/10.1007/978-3-031-62355-4

Translation from the Spanish language edition: "Teoria de Colisiones Vehiculares" by Luis Gonzalo Mejía Cañas,
© Luis Gonzalo Mejia Ph.D., M.A., M.Sc. 2019. Published by Luis Gonzalo Mejia Ph.D., M.A., M.Sc. All Rights
Reserved.

This Springer imprint is published by the registered company Springer Nature Switzerland AG
The registered company address is: Gewerbestrasse 11, 6330 Cham, Switzerland

If disposing of this product, please recycle the paper.

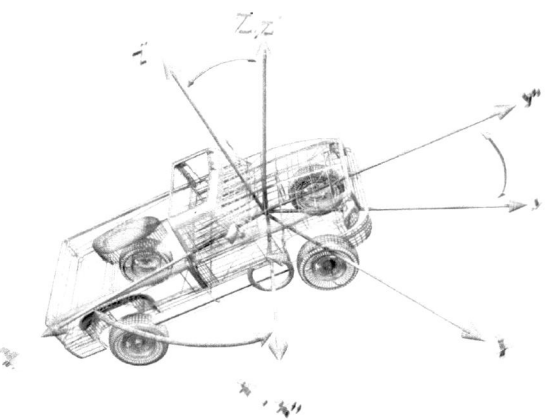

.

My parents, my wife Gloria Isabel
and our children
Carlos Federico and Maria Isabel.

Introduction

As part of physics, mechanics deals with forces and their effects, and in turn is divided into dynamics, which deals with the moving bodies, and static, which deal with bodies at rest, i.e., those whose movement is zero and represent a special case of dynamics. The dynamics are divided into kinematics which deals with the space-time description of the movement of a body without considering the forces that produce it and kinetics which studies the effects of forces on bodies. These notes will use the term dynamic interchangeably as the consideration of the indicated classification is of no importance in the study of vehicle collisions.

It is important to mention that physical laws in their clarity and simplicity represent an abstraction of the complex nature of things, and that is why each system, each phenomenon, must be reduced to a model that allows the application of such laws. In each problem, the model can be refined according to the requirements of the situation being studied, but it should not be forgotten that a simple model accompanied by good judgment is preferable, then a complex one coupled with a total ignorance of the phenomenon and an absolute lack of common sense.

The phenomena present in a vehicle collision are extremely complex to such an extent that car manufacturers are forced to carry out full-scale tests in order to draw conclusions about vehicle behavior. Fortunately, for the engineering calculations required in a collision analysis, simple equations can be considered, which, with reasonable development, and as already said, with good judgment, lead to reliable results.

This book seeks to take a first step on the exciting topic of the Theory of Vehicular Collisions because this field of science has a crucial importance in clarifying circumstances that can lead to situations with serious criminal and civil repercussions. Finally, it is necessary to mention that, in this book, the word vehicle refers to any means of transport, be it a bus, a car, a motorbike, or a bicycle, as the laws that govern the dynamics of movement apply equally to all these means of transport.

I thank my son Carlos Federico Mejía for his dedicated work in the preparation of most of the graphics of this book (Fig. 1).

Fig. 1 (left) and (right): to start developing good judgment, it is very interesting to look at these two photos and try to conclude what really happened, that is, which car was responsible for the collision. Taken and duplicated from El Colombiano Newspaper August 16, 2019, to show a real important situation in collisions research

Contents

The Scientific Method in the Theory of Collisions

Until the thirteenth century the philosophical *Doctrine of Apriorism* was applied without contemplation, and if the result of any experiment or observation conflicts with an accepted theory, the experiment or observation had to be bad or was the product of a diabolical influence. Thus, for example, it was supposed, a priori, that the sun was turning around the earth and that all the planets were moving in circular paths, because was reasonable and compatible with accepted dogmas that this happened, even if this would find many inaccuracies and difficulties with the determination of trajectories and planetary movements.

Nowadays apriorism is not accepted in research, thanks to the work of monk Francis Bacon who gave the basis of the *Scientific Method* used today, that points out that knowledge is obtained through a joint work of: observations, experiments and good judgment.

The Scientific Method proceeds as follows:

1. Initially, detailed observations of actual events are made.
2. A *working hypothesis* is formulated to explain those observations (Fig. 1.1).
3. Additional experiments and observations are made to test the predictive ability of the working hypothesis.
4. With more observations and judgement, sometimes may be necessary to modify or discard the original hypothesis.

In order to make other observations useful, it is very important to try to make all reports from different sources comparable, for example by using a uniform accident description (see Chap. 12).

L. G. Mejía Cañas, *Introduction to the Theory of Vehicular Collisions*, Synthesis Lectures on Mechanical Engineering, https://doi.org/10.1007/978-3-031-62355-4_1

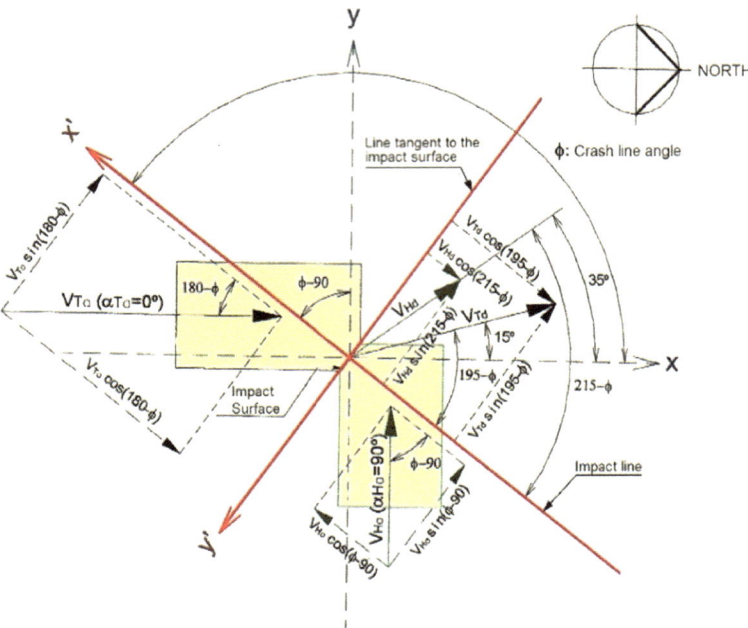

Fig. 1.1 First stage of the study of a working hypothesis in the scientific method

At this point it is very valuable to quote the researcher Noon [1] directly.

"The scientific method is directly applicable to vehicular accident reconstruction. Conclusions about an accident should be consistent with the whole body of available physical evidence and verifiable observations, and should be consistent with proven scientific principles."

"The doctrine of apriorism, unfortunately, is often applied to accident reconstruction work, by a number of poor practitioners. This occurs when the investigator has decided ahead, of time what happened, and simply looks for a few facts to support his or her initial assumptions. It is worth remembering that people that those who believe the earth is flat can also point to a small number of observations to prove their claim."

The effect of a collision was known since ancient times. Aristotle (384-322 BC) in his *Questiones Mechanicae* recognized the difference between the pressure exerted on a still body and the effect of the collision with an object moving very fast.

Then followed centuries of research in which it was necessary to distinguish the concepts of mass and weight, because, as Professor Mach [3] points out, Galileo always thought that these two concepts were the same and geniuses like Huygens in his works always wrote "corpus majus" (big body) and "corpus minus" (small body).

Galileo (1564–1642) in his *Discorsi* quantitatively analyzed the problem of collisions, but it was Marcus Marci (1595–1667) who first quantitatively addressed the problem of elastic collisions and in 1639 published his work *De Proportione Motus,* see Fig. 2.1a [2] in which he analyzed different aspects of movement and collisions. The Fig. 2.1b [2] shows one of Marci's experiments related to this matter.

Years later, Descartes (1596–1650) published in Amsterdam, his *Philosophiae Principal* in which he established 7 rules concerning the collision of two perfectly hard bodies *(perfect dure).* Responding to a call from the Royal London Society, in 1668 mathematician John Wallis, master Christoper Wren and physicist Christian Huygens presented different solutions to the problems of body-to-body collisions.

Isaac Newton (1643–1727) advanced Huygens' research essay, finding that, for theory and experiments to coincide, it was necessary to consider that the bodies were not perfectly elastic and defined the so-called "restitution coefficient e", that is worth 0 for an inelastic collision and 1 for an elastic collision. Figure 2.2a [2] published in his *Pilosophiae Naturalis Principia Mathematica,* (1687) shows the device used by Newton to measure the restitution coefficient. This coefficient is of enormous importance in the study of collisions as will be seen below.

© The Author(s), under exclusive license to Springer Nature Switzerland AG 2025   3
L. G. Mejía Cañas, *Introduction to the Theory of Vehicular Collisions*, Synthesis Lectures on Mechanical Engineering, https://doi.org/10.1007/978-3-031-62355-4_2

(a)

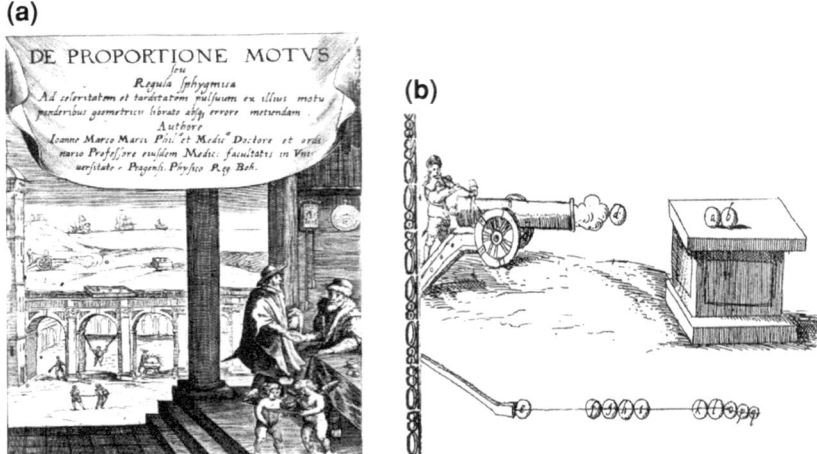

Fig. 2.1 **a** Marcus Marci: De Proportione Motus. **b** Marcus Marci's collisions experiments

Fig. 2.2 **a** Isaac Newton, his main and monumental work and the device that he employ to measure the restitution coefficient "e" used in collision investigations. **b** Leonard Euler [2]

(a)

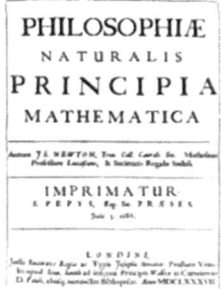

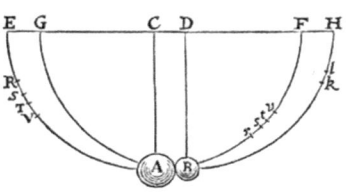

Fig. 2.2 (continued)

(b)

Finally, Leonhard Euler (1707–1783) extended these investigations to two-dimensional collisions, whether they are central or eccentric collisions (see Chap. 7). He was the first to propose the "Impulse Principle and the amount of movement" which is why he shares with Newton his importance in the development of dynamics.

Equally they are due to him the "Principle of Moments", Both principles are of enormous importance in the study of collisions.

With this brief historical overview, it is possible to conclude, that the basis for the study of collision theory has been given for hundreds of years. As we will see in Chap. 12, the advent of computers has facilitated the study of the multiple situations that may arise in a collision investigation.

From Agatha Christie's Poirot, to Reflect and Conclude:

Jewel Robbery at the Grand Metropolitan
How did Poirot discover this? As always, Monsieur, it's the little details, inconsequential things, which attracts Poirot's attention.

Tragedy in Three Acts
1. Well, what do we do now?: Think.
....
2. I needed to prove to myself something that reason dictates to me.
....
3. That was obvious. However, that obvious was a manipulated obviousness. I mean, it was too obvious and therefore nothing obvious.

The following four methodologies are primarily used in collision analysis: the energy method, the momentum or amount of motion method, the dynamic method, and the kinematic method.

The energy method
Makes use of the first law of thermodynamics, which states that the total energy of a system before a collision is equal to the energy of the system after the collision plus the irreversible work performed in the interval of the accident. The above can be expressed by means of the following equation:

$$E_i = E_f + U, \text{being} :$$

E_i: The total energy of the system, kinetic and potential, before the collision,

E_f: The total energy of the system, kinetics and potential, after the collision and,

U: The amount of irreversible work performed during the collision process, due to braking, friction, overturning, shocks, etc.

The momentum method
Applies the principle of conserving the amount of movement, equalizing the momentum of the vehicles before and after the collision, allowing the determination of the initial speeds. The fact that collisions between vehicles are usually semi-plastic, i.e., they are not fully elastic or totally plastic, can be considered using the so-called restitution coefficient "e" (see Chaps. 2 and 8), which is a measure of energy lost during impact. The restitution coefficient "e" is worth 1 for a fully elastic shock, a situation in which is accepted the hypothesis that no permanent energy loss occurs during the collision and is worth 0 for a

L. G. Mejía Cañas, *Introduction to the Theory of Vehicular Collisions*, Synthesis Lectures on Mechanical Engineering, https://doi.org/10.1007/978-3-031-62355-4_3

completely plastic shock, in which crashing vehicles stick together and continue to move with the same speed.

The dynamic method
Applies the laws of Newton's movement, seeking to find the initial speeds of the vehicles, but it requires a large number of detailed calculations and often needs to assume forces and accelerations, until it achieves a result that conforms to the parameters found in the accident.

The kinematic method
Adopts some behavioral parameters of vehicles and uses motion equations in their application.

Of these four methods, the momentum method is the most commonly used and will be considered in the development of these book. In crash investigations, two or more techniques are usually used to analyze the convergence of the results and, as the energy method is a powerful tool for determining speeds before impact, it is usually used as a complement to the momentum conservation method.

This book includes an easy to use but powerful author's program called "Two dimensional dynamics of collisions" which was developed using the above-mentioned methodologies. In Chap. 12, we will deepen about these issues.

Introduction to Vehicular Dynamic

4

Although it was Galileo (1564–1642) who made the first contributions to the dynamics, it was Isaac Newton (1643–1727) who published in 1687, one of the high works of human thought, his *Pilosophiae Naturalis Principia Mathematica* in which he laid the foundations of classical mechanics and proclaimed the fundamental laws of the movement [2–4]:

Law I: Everybody continues in its state of rest or uniform movement in a straight line unless it is forced to change that state by forces acting upon it.

Law II: The change in the amount of movement is proportional to the force that has been given and follows its direction.

Law III: An equal reaction is always opposed to every action i.e., the reciprocal actions of two bodies on each other are always equal and directed to opposite directions.

In bodies moving at speeds of tens or even hundreds of meters per second, such as motor vehicles or airplanes, classical mechanics is fully applicable, with the above laws being the basis for developing the theory of vehicle collisions. It is important to mention that, for example, with the help of those laws, specifically the first law and the injuries that passengers receive from a vehicle, allow to determine in which direction the vehicles were going before the collision. Thus, for example, in a crash at an intersection where the vehicle coming from the right (#1) hits another vehicle sideways (#2), it is expected that the driver and passengers of the collided vehicle suffer severe injuries on their left side and in turn, those who caused the collision, receive them frontally. Injuries would be different if vehicle 2 collided with vehicle 1, as shown in Fig. 4.1.

© The Author(s), under exclusive license to Springer Nature Switzerland AG 2025   9
L. G. Mejía Cañas, *Introduction to the Theory of Vehicular Collisions*, Synthesis Lectures on Mechanical Engineering, https://doi.org/10.1007/978-3-031-62355-4_4

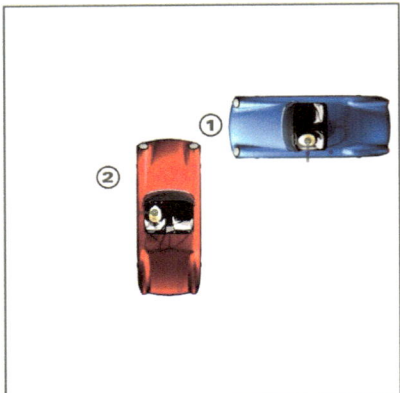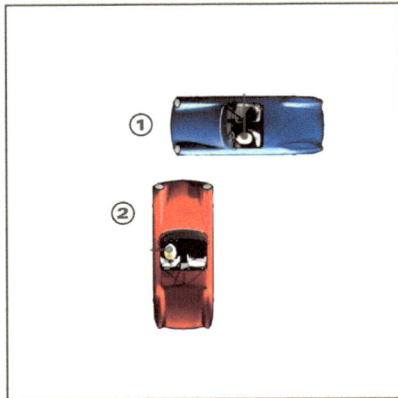

Fig. 4.1 Types of injuries depending on the type of collision

The second and third laws allow to propose the *method of momentum* or quantity of movement, one of the methods of analysis mentioned in Chap. 3, which will be developed in Chap. 6.

Concept of the Center of Mass

The center of mass is the point where the mass of a body is considered to be concentrated and therefore, the coordinates and of the center of mass for a body with masses m_1, m_2 … m_i with coordinates (x_i, y_i) are defined as:

$$\bar{x} = \frac{\sum\limits_i m_i x_i}{\sum\limits_i m_i} \quad \bar{y} = \frac{\sum\limits_i m_i y_i}{\sum\limits_i m_i}$$

The center of mass has the following very important property that when is acted by external forces, moves just as if all the masses were concentrated at that point and were acted by a resulting force equal to the sum of the external forces applied on the system.

This property allows to describe the movement of rigid bodies as a combination of the translation of its center of mass and a rotational movement around an axis passing through the center of mass, as indicated in Fig. 5.1, during the event (left figure) and at the final position (right figure). Note that the center of mass moves in a straight line, an important fact that is valid for vehicle collision analysis.

An exception of the previous conclusion occurs when all the three moments of inertia are different and can be presented a coupling of rotational movements on the three main axes, producing a nonlinear displacement [1].

As we noted earlier in Chap. 3, sometimes the kinetic energy may be required in crash analysis. In this case, from the speed of translation of the center of mass of a vehicle, as well as its angular velocity, it is possible to find the kinetic energy of the vehicle.

To achieve it, it will then be considered a body of mass m, which moves in a straight line and at the same time rotates with angular velocity ω. The speed of a typical point of the body v_i, with mass m_i is the sum vector of the velocity v_{cm} of center of mass and the

© The Author(s), under exclusive license to Springer Nature Switzerland AG 2025   11
L. G. Mejía Cañas, *Introduction to the Theory of Vehicular Collisions*, Synthesis Lectures on Mechanical Engineering, https://doi.org/10.1007/978-3-031-62355-4_5

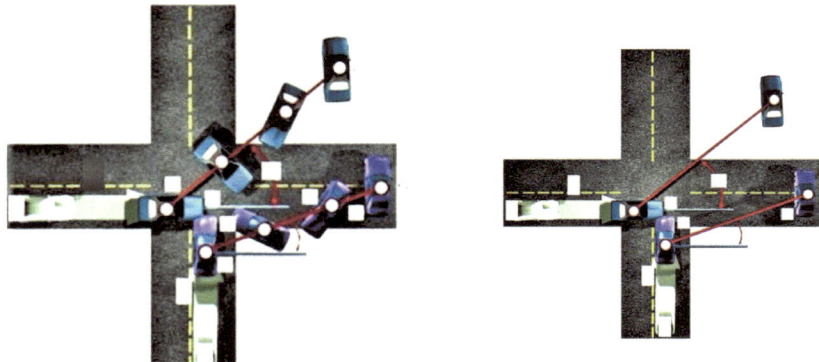

Fig. 5.1 Displacement of a body that translate with rotation

translation speed vi' of the point in question, relating to the center of mass:

$$\vec{v_i} = \vec{v_{cm}} + \vec{v_i'}$$

In this way, the kinetic energy K_i of the particle m_i is:

$$K_i = \frac{1}{2}m_i v_i^2 = \frac{1}{2}m_i\left[\vec{v_i} \cdot \vec{v_i}\right]$$

$$K_i = \frac{1}{2}m_i\left[\vec{v_{cm}} + \vec{v_i'}\right] \cdot \left[\vec{v_{cm}} + \vec{v_i'}\right]$$

$$K_i = \frac{1}{2}m_i v_{cm}^2 + m_i\vec{v_{cm}} \cdot \vec{v_i'} + \frac{1}{2}m_i v_i'^2$$

And the kinetic energy of the entire body is:

$$K = \sum K_i = \sum \frac{1}{2}m_i v_{cm}^2 + \sum m_i\vec{v_{cm}} \cdot \vec{v_i'} + \sum \frac{1}{2}m_i v_i'^2$$

$$K = \frac{1}{2}\left(\sum m_i\right)v_{cm}^2 + \vec{v_{cm}} \cdot \sum m_i \vec{v_i'} + \sum \frac{1}{2}m_i v_i'^2$$

Now how:

- $\sum m_i$ equals the total mass of the body.
- $\sum m_i v_i'$ is m times the velocity of the center of mass with respect to the center of mass, zero.i.e.
- $\sum \frac{1}{2}m_i v_i'^2 = \frac{1}{2}I_{cm}\omega^2$

Then:

$$K = \frac{1}{2}mv_{cm}^2 + \frac{1}{2}I_{cm}\omega^2 \tag{5.1}$$

As can be seen in the above equation, the total kinetic energy of a rigid body that moves in a straight line and at the same time rotates, has a component $\frac{1}{2}mv_{cm}^2$ corresponding to translation of the center of mass and another one $\frac{1}{2}I_{cm}\omega^2$ which corresponds to the rotation around an axis passing through the center of mass.

The Momentum Method

Newton's second law can be expressed mathematically as $F = ma$ or $F = d(mv)/dt$ and as the mass remains constant as:

$$F = md(v)/dt \tag{6.1}$$

Integrating it between times t_1 and t_2 and rearranging we obtain:

$$mv_1 + \int_{t_1}^{t_2} F dt = mv_2 \tag{6.2}$$

Then it is possible to conclude that the Eq. (6.1) may be written too as a function of the change in momentums through Eq. 6.2. In this equation v_1 and v_2 are the speeds of the center of mass in times t_1 and t_2 respectively. The integral is know as linear impulse and the product mv as the amount of linear motion. Both quantities are vectors.

This equation is extremely important in collision dynamics and is known as the "*Principle of Impulse and amount of Movement*" and in words, it simply says that the impulse of a force is equal to the change in the amount of movement of the body on which it acts.

As can be seen, if on a vehicle traveling in one direction, indicated by the angle α_1 a force F acts in a time interval $t_2 - t_1$ the effect this force on this vehicle is to modify its direction and its amount of movement, changing the speed from mv_1 *to* mv_2. Although this fact seems obviously, it took centuries for humanity to find the mathematical expression of this phenomenon. Finally, it is important to note that the impulse, $\int_{t_1}^{t_2} F dt$ can be produced by another vehicle or by a rigid or flexible barrier.

© The Author(s), under exclusive license to Springer Nature Switzerland AG 2025
L. G. Mejía Cañas, *Introduction to the Theory of Vehicular Collisions*, Synthesis Lectures on Mechanical Engineering, https://doi.org/10.1007/978-3-031-62355-4_6

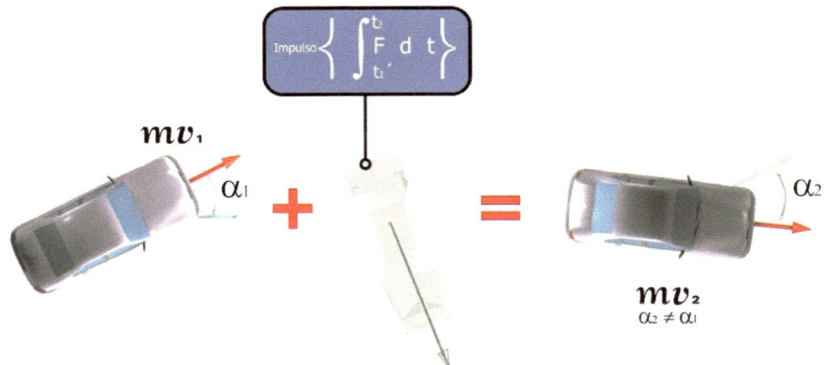

Fig. 6.1 Graphical meaning of the "*Principle of Impulse and amount of Movement*"

Before proceeding, it is advisable to analyze with an example, the concept of impulse and amount of movement *mv* conclusions that seem to be surprising can be reached: for example, a person with a mass of 70 kg who rides a bicycle at 15 m/s has more amount of movement $70 \times 15 = 1050$ kg-m/s than a 1-ton car traveling at 1 m/s, whose amount of movement is $1000 \times 1.0 = 1000$ kg-m/s.

As a practical conclusion for the case of vehicle collisions that we are dealing with, is that a vehicle of small mass but traveling at a high speed has more amount of movement than a higher mass that travels at low speed, so the smaller vehicle could cause, greater damage to the larger vehicle, as can be seen in the Fig. 6.2.

*Top understand this point is vitally important for cyclists, skateboarders and scooters users, between others, who often drive at high speeds in pedestrian areas, **putting at risk the lives of pedestrians** that walk quietly in them.*

It is equally interesting to imagine the following test: a vehicle with *m* mass travels at a *v* speed and crashes into a wall (Fig. 6.3).

If the wall is rigid, the change in the amount of movement from *mv* to 0 occurs in a very small-time t and therefore the impact force F is very large. Conversely, if springs (which can be rims) are arranged on the wall, the change in the amount of movement occurs in a very large time Interval T and therefore the impact force f is small.

In conclusion, the vehicle and the occupants, in the case of rigid barrier, receives a greater force than in the case of flexible barrier and would therefore suffer more severe damage.

Example

Let's consider a vehicle that is going at a speed of 60 km/h, that is at approximately 16.7 m/s and collides with a rigid barrier that immediately stops the vehicle.

Fig. 6.2 A real situation that illustrate the importance of the amount of movement

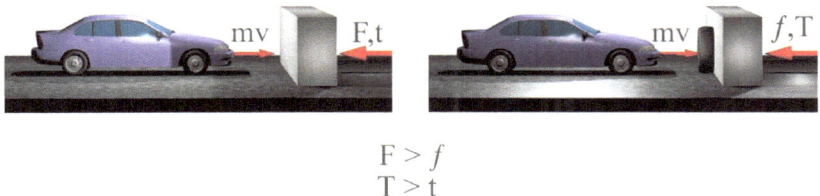

$$F > f$$
$$T > t$$

Fig. 6.3 The force **F** applied to the vehicle in the case of a rigid barrier is greater than the force **f** that it receives If he were to crash with a flexible barrier. On the other hand, the time of the collision in the case of the rigid barrier **t** is less than that of the collision **T** against the flexible barrier

In mathematical terms, we could say that the process was carried out in a very short time, for example, one hundredth of a second (0.01 s) and, as a result, there is subjected to a deceleration of $16.7/0.01 = 1670$ m/s^2, leaving the driver subject to an equivalent force 170 times the force of gravity, being completely shattered, with no chance of survival.

Otherwise, if the barrier were flexible, the impact time would be much longer, for example 1 s and in this situation, de deceleration will be of 16.7 m/s^2, having the occupants a chance to survive.

Once these concepts are defined, it is possible to begin with the study of the types of vehicular collisions.

Historical Note:

Christiaan Huygens and his enormous contribution to the theory of collisions.

Regulæ de Motu Corporum ex mutuo impulfu.

1. *Si Corpori quiefcenti duro aliud æquale Corpus durum occurrat, poft contactum hoc quidem quiefcet, quiefcenti vero acquiretur eadem qua fuit in Impellente celeritas.*

2. *At fi alterum illud Corpus æquale etiam moveatur, fraturque in eadem linea recta, poft contactum permutatis invicem celeritatibus ferentur.*

3. *Corpus quamlibet magnum à corpore quamlibet exiguo et qualicunque celeritate impacto movetur.*

4. *Regula generalis determinandi motum, quem corpora dura per occurfum fuum directum acquirunt, hæc eft:*
Sint Corpora A et B, quorum A moveatur celeritate A D, B vero ipfi occurrat, vel in eandem partem moveatur celeritate B D..

(928)

B D, vel denique quiefcat, hoc eft, cadat in hoc cafu punctum in B. Divifâ lineâ A B in C, (centro gravitatis Corporum A B.) fumatur C E æqualis C D. Dico, E A habebit celeritatem corporis A poft occurfum; E B vero, corporis B; et utrumque in eam partem, quam demonftrat Ordo punctorum E A, E B. Quodfi E incidat in punctum A vel B, ad quietem redigentur corpora A vel B.

5. *Quantitas motus duorum Corporum augeri minuive poteft per eorum occurfum; at femper ibi remanet eadem quantitas verfus eandem partem, ablatâ inde quantitate motus contrarii.*

6. *Summa Productorum factorum à mole cujuflibet corporis duri, ducta in Quadratum fua Celeritatis, eadem femper eft ante et poft occurfum eorum.*

7. *Corpus durum quiefcens, accipiet plus motus ab alio corpore duro, fe majori minorive, per alicujus tertii, quod media fuerit quantitatis, interpofitionem, quam fi percuffum ab eo fuiffet immediatè. Et fi corpus illud interpofitum, fuerit medium proportionale inter duo reliqua, fortius aget in quiefcens.*

Christiaan Huygens (1629–1695) and its 7 rules about collisions theory, published in 1669 [2].

The crash line **x´** is defined as the normal to the contact surface **y´** of the colliding vehicles (Fig. 7.1).

Based on this definition of the crash line, the following types of collisions can be distinguished.

7.1 Central Collisions

Is one in which the crash line x´ passes through the mass centers of colliding vehicles. If the direction of the vehicles is the same as the crash line, we have a direct central collision (Fig. 7.2a) and when one or both vehicles move along a direction different from the crash line before the collision, we have an oblique central collision (Fig. 7.2b). The direct central collision is, therefore, a particular case of the oblique central collision.

7.2 Eccentric Collisions

Occurs when the crash line x´ does not pass through the mass centers of vehicles (Fig. 7.2c).

This publication is not intended to be exhaustive in the development of formulas for all types of collisions and we will simply develop, in the case of an oblique central collision, the equations describing the linear movement of vehicles, using for the effect of the "Principle of impulse and amount of movement". Initially, the rotation of vehicles will not be considered, as it is widely known that after an eccentric collision, that is,

L. G. Mejía Cañas, *Introduction to the Theory of Vehicular Collisions*, Synthesis Lectures on Mechanical Engineering, https://doi.org/10.1007/978-3-031-62355-4_7

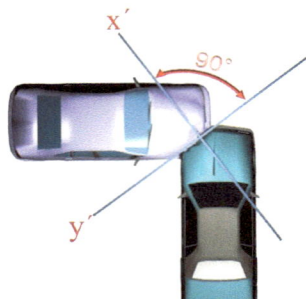

Fig. 7.1 Definition of the crash line

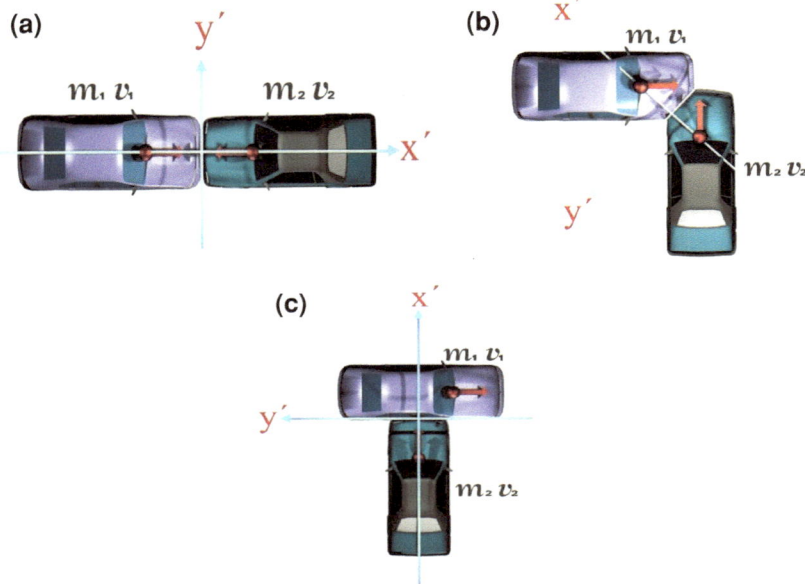

Fig. 7.2 a Direct central collision. **b** Oblique central collision

in which rotations occur, the center of mass moves over a straight trajectory as shown in Figure 5A, corresponding to the movement of a top (spin). This important natural law had been discovered by physicist Christian Huygens in the 17th century, who noticed, and proved it for spherical bodies, that in central or oblique collisions of two or more bodies, the center of mass before and after the collision moves in a straight line with equal speed [2]. Anyone interested in deepening in this issue, can refer to references [8, 9 and 10].

Historical Note:

Pioneering research from Galileo Galilei (1564–1642)

His work in the field of materials resistance has prepared the way for the development of the structural theories, including the collision theory. He questioned all existing knowledge and gave a great boost to the scientific method.

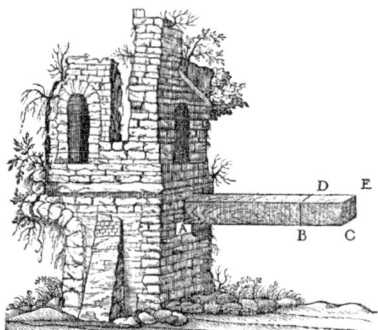

This figure, related with their test on beams, appear in his book *Discorsi e dimonstrazioni matematiche, intorno a due nuove scienze*. Leyden 1638. [2]

Coefficient of Restitution "e"

Before starting with the development of equations, it is convenient to define the concept of the "Coefficient of Restitution", which plays a fundamental role in collision theory. As already mentioned in Chap. 2 (Fig. 3), it was Newton who first defined this concept.

The *Coefficient of Restitution* is the relationship between $\int Rdt$ of the impulse of force R (force exerted by one body to the other during the period of recovery of the collision) and the value $\int Pdt$ of the impulse of force P (force exerted by one body to the other during the period of the deformation of the collision), and can be found by the following expression (Fig. 8.1):

$$e = |\text{Relative separation speed axis } x' / \text{ relative approach speed axis } x'| \qquad (8.1)$$

Due to the above definition, the restitution coefficient **e** is applied in the direction of the crash line, it is where the forces R and P occur. If the collision is a "Oblique central Collision" (Fig. 7.2) is necessary to separate the component vectors (§ 15.1).

For any crash, $\int \boldsymbol{Pdt} \geq \int \boldsymbol{Rdt}$, and for this reason $0 \leq e \leq 1$. When $e = 0$ the crash is called plastic and when $e = 1$ it's called elastic. Since the restitution coefficient only acts in the direction of the crash line, then, if this is zero and there is also no speed component in the direction tangent to the crash (as in a direct central collision) the vehicles continue together (glued) after the collision. In any other case the vehicles "bounce", the greater the separation between them as the closer to 1 is the restitution coefficient.

It's important to note that this coefficient depends on the speeds of the vehicles, the materials and their shape [8] and their determination for each case, should be done experimentally. Large automakers determine, among others, this value in collision testing with real vehicles.

L. G. Mejía Cañas, *Introduction to the Theory of Vehicular Collisions*, Synthesis Lectures on Mechanical Engineering, https://doi.org/10.1007/978-3-031-62355-4_8

Fig. 8.1 **x´** is the crash line, the impact surface is marked in red, **y´** is tangent to the impact surface and **Ø** is the crash line angle

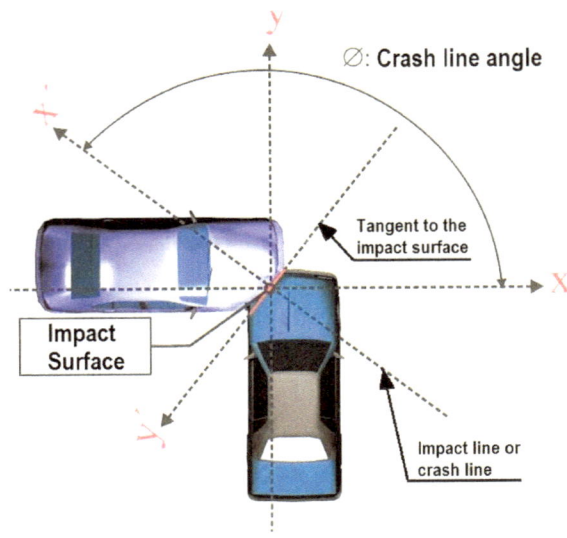

8.1 Restitution Coefficient for a Semiplastic Collision

When two bodies collide, some of the kinetic energy is dissipated into plastic deformation. This energy is given by the following expression [1]:

$$E_{dis} = \frac{m_1 m_2 (V_{i1} - V_{i2})^2}{2(m_1 + m_2)} (1 - e^2)$$

(8.1)

So, when the crash is totally plastic, e = 0:

$$E_{dis(ch. plastic)} = \frac{m_1 m_2 (V_{i1} - V_{i2})^2}{2(m_1 + m_2)}$$

(8.3)

and when the collision is totally elastic, e = 1:

$$E_{dis(elastic)} = 0$$

and then when the collision is semi-plastic, the dissipated energy is:

$$E_{dis(semi plastic)} = \frac{m_1 m_2 (V_{i1} - V_{i2})^2}{4(m_1 + m_2)}$$

(8.4)

and the restitution coefficient for this type of collision is obtained replacing the latter value in the general equation for energy dissipation:

$$\frac{m_1 m_2 (V_{i1} - V_{i2})^2}{4(m_1 + m_2)} = \frac{m_1 m_2 (V_{i1} - V_{i2})^2}{2(m_1 + m_2)} \left(1 - e^2\right)$$

$$1 - e^2 = \frac{1}{2}, \textit{from which it can be concluded that } e = \sqrt{\frac{1}{2}} = 0.707, \qquad (8.5)$$

In the Fig. 8.2, the axis x, correspond to the restitution coefficient and the y axis to the dissipated energy E in the collision. If E = 0 the collision is totally elastic, if E = 50% the collision is semiplastic and if E = 100%, the collision is totally plastic.

Fig. 8.2 Relationship between the energy dissipated in a collision and the restitution coefficient

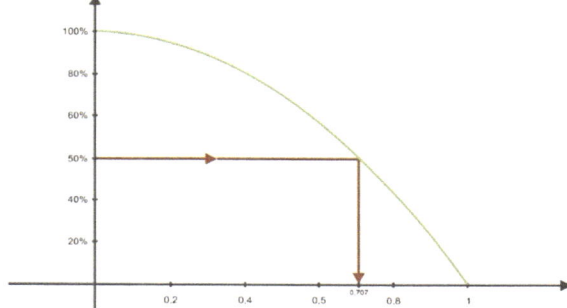

Stages of a Collision

Once the types of collisions have been described, it is advisable to analyze what happens at the time of impact. In principle, a vehicle collision can be separated into two stages, as indicated in Fig. 9.1, for the simplest case of a vehicle colliding with a rigid barrier. The collision starts with an initial contact and then comes the final contact with only translation theoretically in the case of a central impact and with translation with rotation for an eccentric impact, because, in the latter case, the left corner is suddenly braked by the obstacle, but the rest of the vehicle continues to move forward, resulting in a typical rotation in this kind of collisions.

When the crash is between two vehicles, in the Fig. 9.2 are shown the first and the final contact. Between these phases, the *maximum engagement* occur and then the collision develop as follows: "From the first contact to the maximum fit, the left corner of vehicle **A** accelerates in the direction of movement of vehicle **B**. At the same time, the front right corner of vehicle **B** accelerates in the direction of movement of vehicle **A,** causing both vehicles to rotate, making the A in clockwise direction towards vehicle **B** and the vehicle **B** rotating in the opposite clockwise direction towards vehicle **A.** This rotation often results in a secondary impact between vehicles" [12].

© The Author(s), under exclusive license to Springer Nature Switzerland AG 2025   27
L. G. Mejía Cañas, *Introduction to the Theory of Vehicular Collisions*, Synthesis Lectures on Mechanical Engineering, https://doi.org/10.1007/978-3-031-62355-4_9

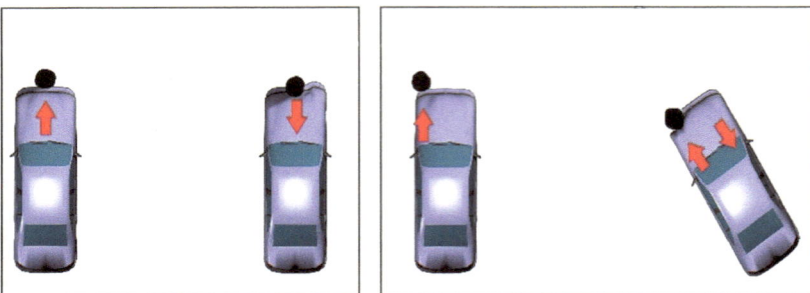

Fig. 9.1 Stages of a collision with a rigid barrier: central impact (left figure) and eccentric impact (right figure)

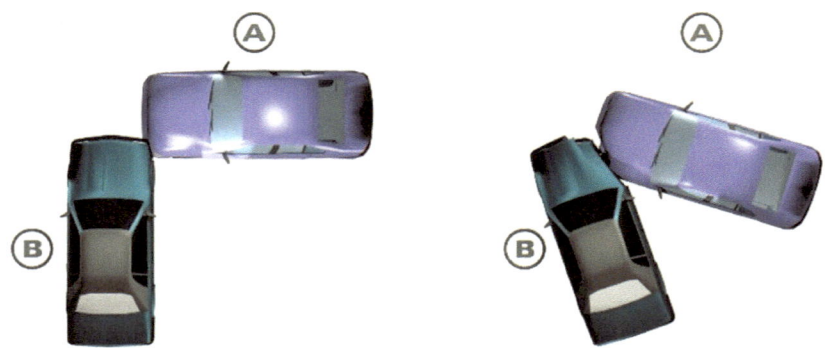

Fig. 9.2 Stages of a collision between two vehicles: first and final contact

Finally, it is important to mention that on average an impact lasts approximately 0.12 s, corresponding 0.06 seconds to the *maximum engagement* [12].

Modelling of a vehicle for an "accurate" analysis of collision is a work that is not yet completely solved, but we the use of the methodologies of "Soft Computing" useful answers are obtained.

The Fig. 10.1 [] show the basic model for the analysis of the vehicle movements. For an analysis closer to reality, the use of the methodology of finite elements and the consideration of several degree of freedom it is mandatory.

It should be noted that the behavior of the springs is nonlinear and therefore, for an accurate analysis the curves $P(t)$ vs $\delta(t)$ must be known.

The condition of the parts of a vehicle also plays an important role in determining the response. Note for example in Fig. 10.2 taken from a Toyota brochure, that the response of a vehicle with shock absorbers, that is with damping (top figure) is completely different from that of one without these (bottom figure). Between these extremes would be the response of a faulty shock absorber.

Modeling a person to study their movement during a collision represents an even more complex task. Figure 10.3 [29, 13] shows:

- A simple model of the human body, that consider resisting forces (springs) and damping forces (viscous damper) and,
- The global coordinate system of the occupants of a vehicle and the coordinate system for placing 9 accelerometers for dynamic studies of the behavior of human body during a collision (standing or seating).

The automotive industry is advancing in the most accurate description of the human body, in order to build ever safer vehicles. In this regard, by way of example, the automobile

L. G. Mejía Cañas, *Introduction to the Theory of Vehicular Collisions*, Synthesis Lectures on Mechanical Engineering, https://doi.org/10.1007/978-3-031-62355-4_10

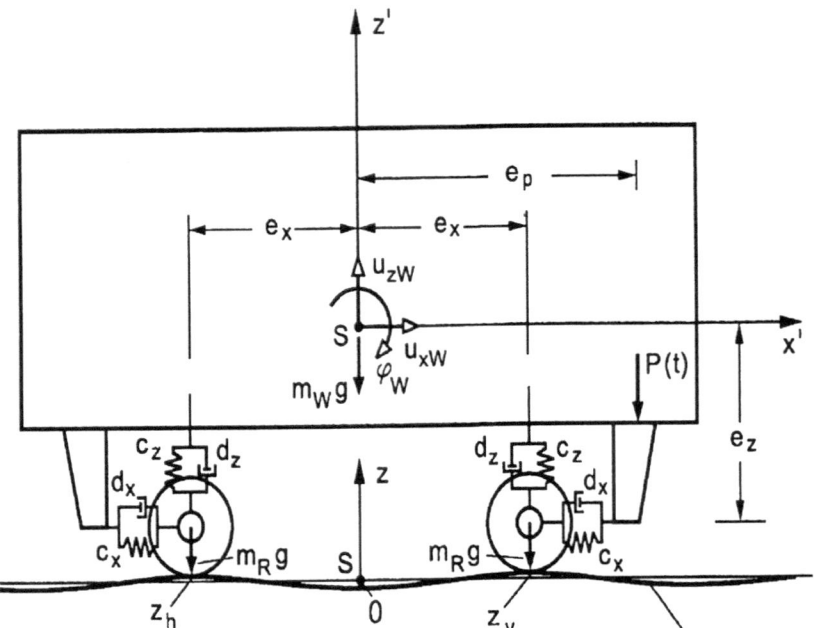

Fig. 10.1 Mathematical model for the dynamic analysis of a vehicle

Fig. 10.2 Vibration of the
body of a vehicle

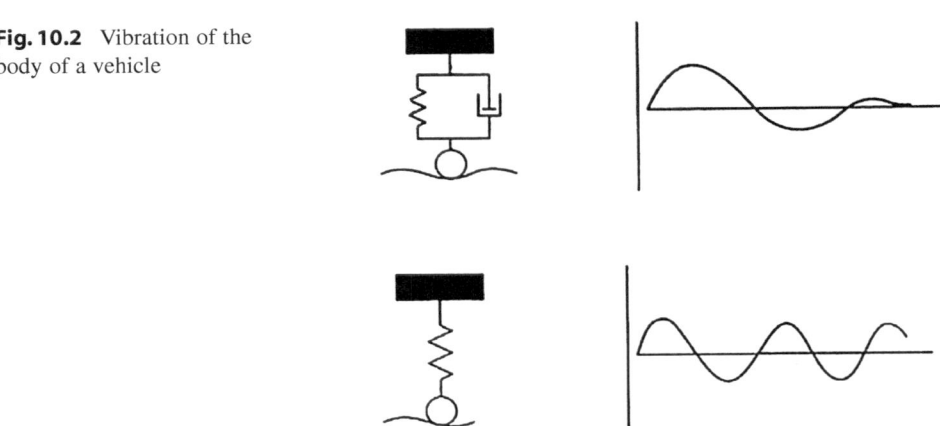

manufacturer Toyota developed years ago the program THUMS (Total Human Model for
Safety) in which, with the help of more than 80,000 cyberparties, it is sought to determine
the injuries that a passenger can suffer in any part of the body (Fig. 10.4).

From the previous discussion, it can be concluded that try to get "exact" answers in
collision problems, with elementary physics formulas, is vain and naive.

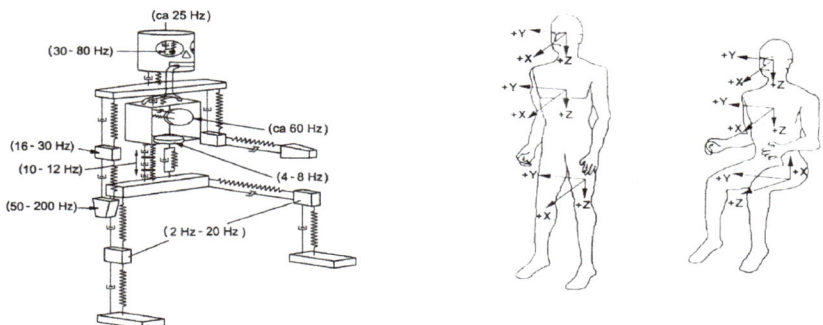

Fig. 10.3 Model of the human body and global coordinate system of a vehicle occupants

Fig. 10.4 Modeling of a
passenger in THUMS

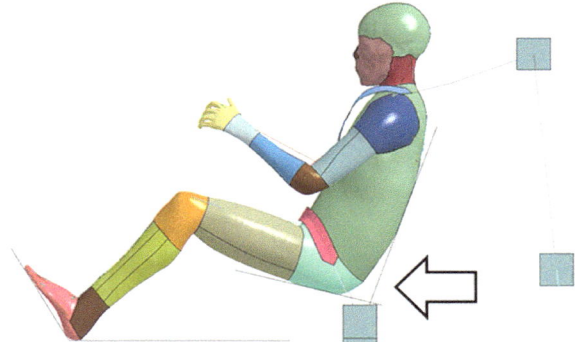

Only inexperienced and non-practical people intend to achieve this, forgetting that the classic analysis of collisions with rigid bodies represents only a first approximation of the problem, which however, with *common sense*, becomes a valuable tool in the search for truth and with the consideration of other evidence it can lead to clarification of the way in which a certain collision occurred.

As in any technical problem, common sense plays a fundamental role in the reconstruction of the facts, even more, in collision problems, in which several of the variables involved are usually unknown. Two sentences of the famous researcher Sherlock Holmes, quoted in Reference 8, express, in a concise and exceptional way, this path of research and reconstruction:

"When the impossible has been eliminated, in what remains, though it seems unlikely, the truth must be" and,

"My name is Sherlock Holmes, and my job is to know what others don't know."

It takes greatness and humility to accept the complexity of a collision problem and first and foremost it is necessary to have the real intention of seeking the truth, otherwise it is of no use to collect evidence and to use the approach of equations and physical models, if you have a preconceived idea and you want, for example, to condemn in advance a person involved in an accident.

The great Master of Engineering, Professor Hardy Cross, in his book "Engineers and Ivory Towers" [28].

wrote:

There has not been a falser point of view than the one that visualizes the engineers inevitably reaching a single solution to their problems through mathematics or laboratory procedures because their solutions are rarely unique. Engineering is not a mathematical science, but it does take advantage of many of the mathematical procedures.

Numbers should be prevented from devouring us and clouding our common sense and good judgment, which, unfortunately every day becomes a sad nightmare, with the complex computer programs available for the collision and other technical analysis, not enough understood by the users and sometimes by the developers,

The German structural engineer Klaus Stiglat masterfully draw in the two attached cartoons [27] this fade from the common sense. And, in the same direction, the Professor Hardy Cross also emphasized two relevant aspects of teaching science [28], that should be considered by the collision's investigator:

- "It is easier to teach rules than to train the good judgment" and
- "A test is worth more than a dozen of expert opinions."

As mentioned above, the approach of physical models and equations must be accompanied by the search for evidence to raise a *first work hypothesis* of how a particular collision occurred. This hypothesis should be analyzed without passion and according to the progress of the investigations, it can be discarded or modified until the answer, that best represents the case under analysis, is reached.

It is not the intention of these notes to describe the multiple techniques of research and reconstruction, however, we will briefly mention some tools that may be helpful in solving crash cases, taking as an example a real collision investigated by the author, which is briefly described below: two cars collide in an intersection, for identification, one was green and the other white (Fig. 12.1), leaving the young driver of the green vehicle dead on site, her friend, who was sitting next to her, quadriplegic and the driver of the white vehicle seriously injured.

12.1 Work Hypothesis

Initially, it is possible to make a model of the collision site to study different possible crash situations.

Once you have obtained a working hypothesis you can use a three-dimensional animation, which allows to "visualize" in three dimensions the adopted working "hypothesis", which must be corroborated or discarded according to physical analyses and the study of

Supplementary Information The online version contains supplementary material available at https://doi.org/10.1007/978-3-031-62355-4_12.

L. G. Mejía Cañas, *Introduction to the Theory of Vehicular Collisions*, Synthesis Lectures on Mechanical Engineering, https://doi.org/10.1007/978-3-031-62355-4_12

Fig. 12.1   A model of the collision site

Fig. 12.2   a Vehicles are approaching to the point of collision. The white vehicle has priority.
b Close up. At this point the driver of the white vehicle can see the green vehicle but can no longer perform any maneuvers to avoid the collision (see paragraph 13.2)

evidence taken at the site, for example, brake prints, traces of paint on vehicles, witness accounts and surveillance camera videos if any, among others.

From the animation prepared for the analysis of this complex night accident, animation that accompanies this book. Four images were taken from this animation and were included (Figs. 12.1, 12.2, 12.3, 12.4 and 12.5) that allow us to have a sequence of what

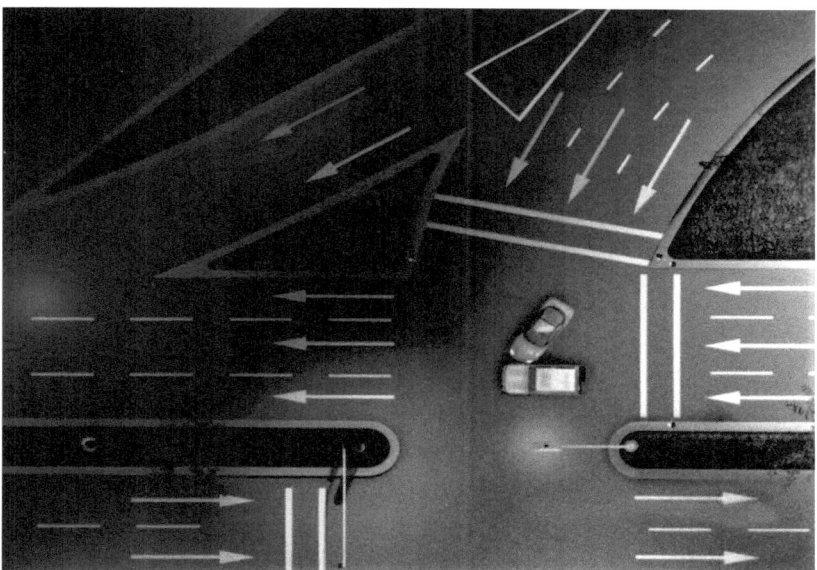

Fig. 12.3 Moment of collision according to the assumed working hypothesis

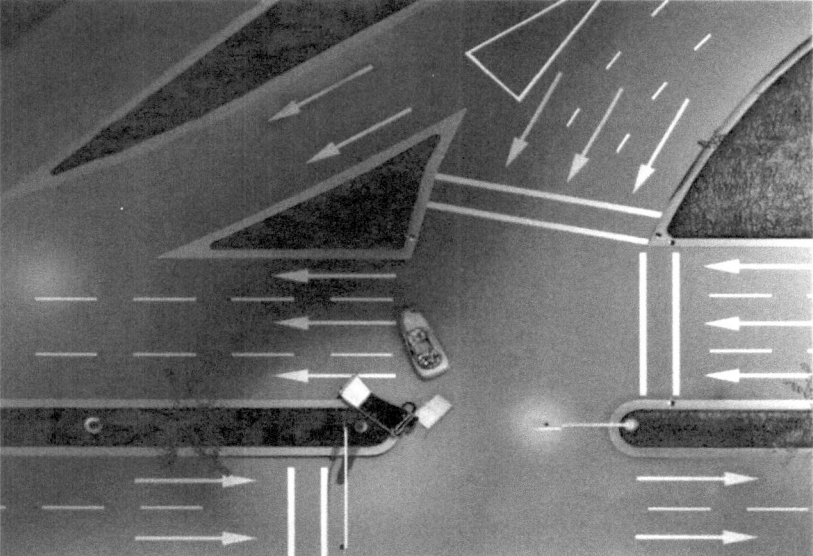

Fig. 12.4 Dynamics of the collision: while the green vehicle move and turns without obstacles, the displacement of the white vehicle is violently stopped by the existence of a rigid concrete pole for surveillance cameras, a situation that produced a second collision in which this vehicle broke and bent at right angles (Fig. 12.7)

Fig. 12.5 This figure shows a fundamental aspect in the analysis of this collision and that is that the green vehicle is lower than the white vehicle and got underneath it, lifting it, making it fly through the air, turning this initially two-dimensional collision into a three-dimensional and multiple one. This aspect of three-dimensional collisions will be discussed in Chap. 17, It is important to mention that in two-dimensional analyses of vehicular collisions it is usually assumed that both vehicles have equal height, that is, that their centers of mass are at the same height from the ground

could have happened in accordance with an adopted working hypothesis. With the help of animation, various situations can be recreated, representing a valuable means for making decisions. Sometimes the use of models or animation can provide a lot of information, including when the roads through which the collided vehicles circulated, have different slopes. The Fig. 12.6 shows the final condition of the crashed vehicles.

From Agatha Christie's Poirot, to Reflect and Conclude:

Murder in Mesopotamia

1. Facts are the cobblestones from which the path we walk is made
2. Why have I been so foolish with how clear the truth is and how simple it is?
3. I'm taking you through the truth

Fig. 12.6 Final condition of the green vehicle after the crash

Fig. 12.7 Final condition of the white car after the multiple tridimensional crashes. As already mentioned, note that the concrete post stopped violently the rotational and translational displacement of the white vehicle and this second collision, practically split the car in two, causing severe secondary damage to the vehicle and serious injuries to the passenger

12.2   Computer Programs: "Two-Dimensional Collision Dynamics"

There are several computer programs that provide invaluable help in vehicle collision studies. One of them, very simple to use is the software "Two-dimensional Collision Dynamics" from the author, developed using the equations contained in this book. This software is extremely useful first of all, because allow to test multiple situations avoiding the complexities that arise when calculations are done by hand. Let's take as an example of application of this software the described collision in Sect. 12.1. Referring to Fig. 12.8:

Input Data

(a) Initial velocities: it is necessary to probe with different initial speed values for each of the two vehicles involved in the collision.
(b) The velocity angles (see Chap. 15).
(c) The masses of the vehicles.
(d) The crash line angle (see Fig. 12.11), taken from the collision sketch prepared by the authorities.
(e) The restitution coefficient: assumed with experience and according to the characteristics of the collision.

With this data, it is necessary to probe until, for an initial velocity determined with common sense and site observations, a situation is found in which the vehicles involved in the collision are left in the positions indicated in the sketch prepared by traffic police authorities. In that drawing, to scale, the results can be drawn by hand can be programmed to animate them.

The process of determining the initial velocities is extremely fast, an aspect that always represents a reason for discussion among those involved in a collision, because each of them usually points out that he was driving at a low speed and that the other driver was going too fast.

Results (Final Data)
So with these few data, the final results, namely the final velocities and angles of velocities are found.

It is important to mention that the analysis of collisions with rigid bodies, considering only their translation, leads to acceptable results, within the framework of accuracy. In this regard we quote Reference 8: "It is almost impossible to determine the number of rotations of a vehicle from the point of impact to the resting point, being therefore impossible to draw the rotation of the vehicle and measure the exact distance that the vehicle rotated in the post-collision phase.

Using as a measurement the distance of the vehicle's center of gravity at the point of impact, to the center of gravity of the vehicle at the rest point, the minimum value of the

Fig. 12.8 Input data and output results of the software "Two-dimensional Collision Dynamics"

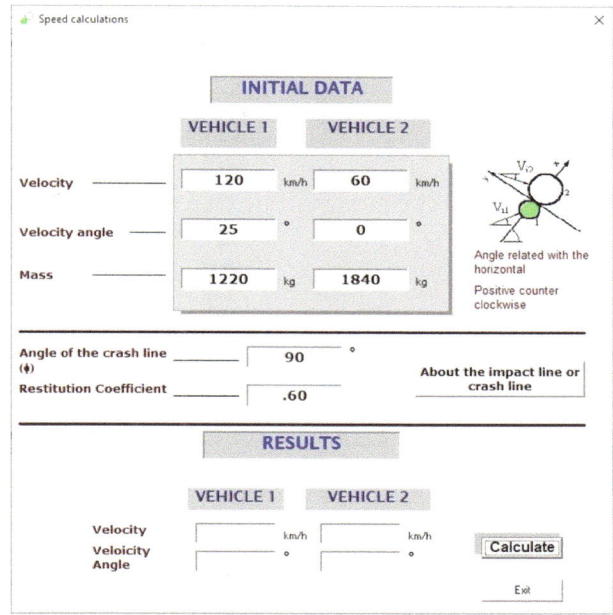

amount of speed and/or energy lost during the post-impact phase can be obtained until it reaches rest."

Recap and Examples

The Fig. 12.9 illustrate the first working hypothesis analyzed with the two-dimensional collision software from author: The three initial lines corresponds to: the first to the preimpact velocities from vehicle #1 (green) and the # 2 (white) respectively; the second to the velocity angles (Chap. 15) and the third to the masses of both cars. Both preimpact velocities and velocity angles are known or assumed with common sense. The fourth line corresponds to the "crash line angle" (Fig. 12.11) and the fifth to the assumed restitution coefficient. The green vehicle is the number 1 (green circle) and the withe vehicle is the number 2 (white circle), which has priority at the street intersection. The black circle in the road separator, correspond to the concrete post indicated in Fig. 12.9 (see Fig. 12.7). As results from software analysis in file sixth the final velocities of the crashed vehicles and in file seventh their corresponding velocity angles are obtained. For clarity, all data and results in the table are drawn on a small plane that appears below the table prepared for each case analyzed.

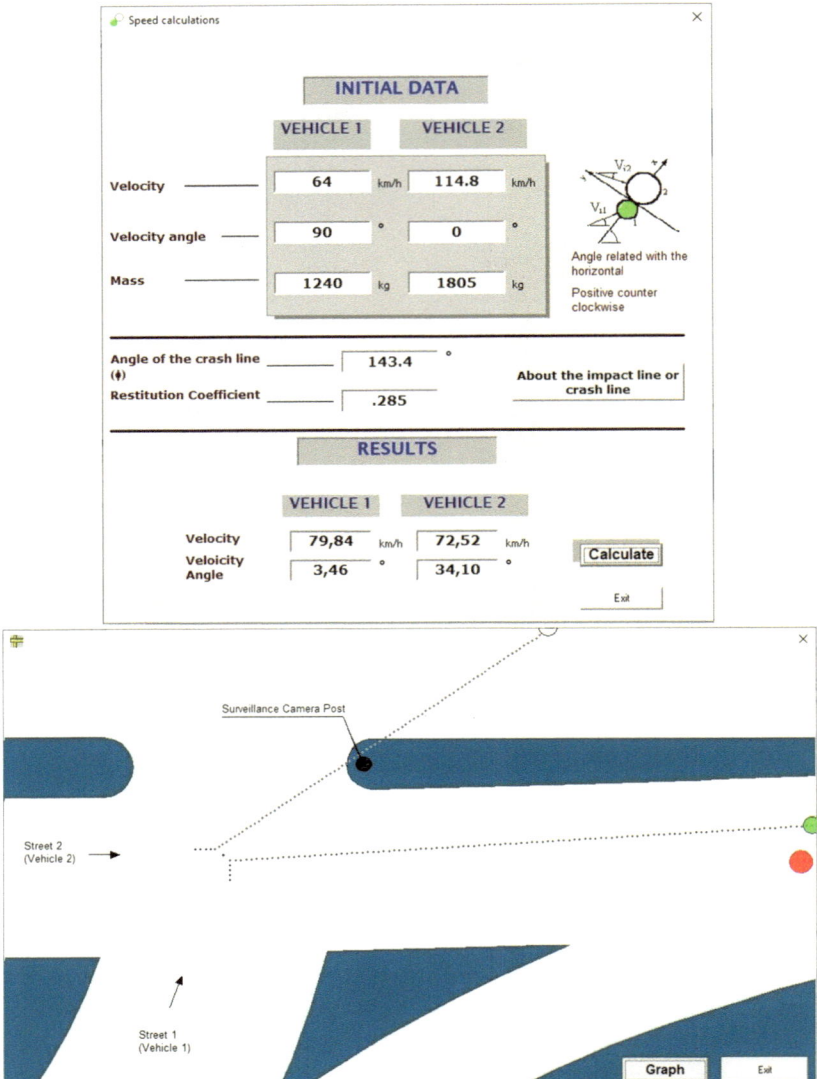

Fig. 12.9 First trial and results including a graphical representation. The black circle corresponds to the surveillance concrete post and in turn, to the final position of the white vehicle. The red circle denotes the final position of the green vehicle. The trajectories of the vehicles (dotted lines) indicate that for the assumed initial data the final position of the collided vehicles will not coincide with those indicated in the police sketch

The trajectories of the vehicles (dotted lines in Fig. 12.9) indicate that for the assumed initial data the final position of the collided vehicles will not coincide with those indicated in the police sketch. Then it is necessary to try again obtaining the final results shown in Fig. 12.10.

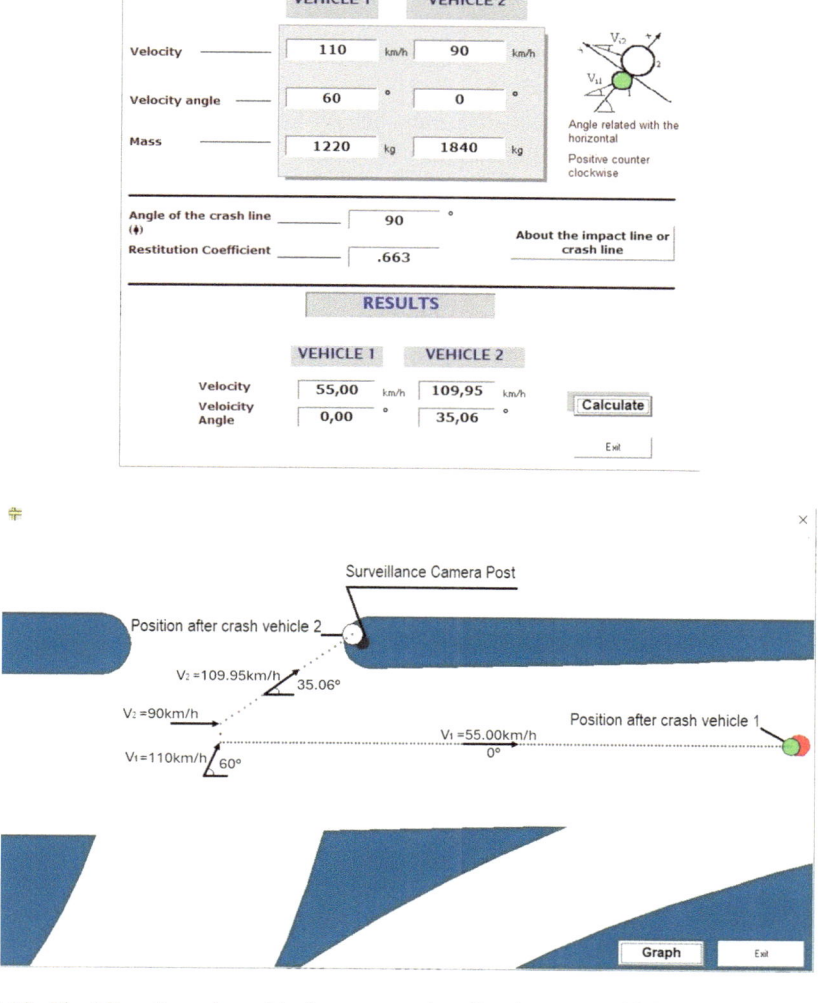

Fig. 12.10 Final Results and graphical representation. For the assumed initial data, the final position of the collided vehicles coincide with those indicated in the police sketch

The following additional facts may be very important during the analysis process:

1. As mentioned, the white car suffered a first collision with the green vehicle and a second one with the concrete post, that is a *multiple crash* occurred. Situations like this, with secondary collisions, greatly complicates the investigation since it is necessary to separate the damages and injures caused in each of these collisions, to reach correct conclusions about what happened in the first collision.

2. The two cars that collides were a green sport car and a white sport-utility vehicle. The green car has a lower height and weight that the white car, an all-terrain car. Despite this very different characteristic, the investigations led to the fact that it was the green car, that did not mark the mandatory stop, was the one who caused the collision. Upon this conclusion, a question immediately arises: how does a small vehicle cause such severe damage to a larger one?

3. Although there are many aspects that must be considered to answer this question, we will mention two by way of illustration:

   (a) From Fig. 12.10, note that the green vehicle traveled at a higher speed, 110 km/h, than the white vehicle that was going at 90 km/h and remembering what was seen in Chap. 6 and Fig. 12.7, what really matters in a collision is the amount of movement i.e. the product mv, which partially explains what happened.

   (b) But that does not clarify everything because which played a fundamental role, was the fact that being the green vehicle lower, it was put underneath the highest white vehicle and leverage it, making it fly, producing a very complex three-dimensional collision. The Fig. 12.5 of the animation allows us to visualize this situation. Later, in Chaps. 16 and 17, an introduction will be made about this type of three-dimensional collisions, which introduce some complexities because it is necessary to use de Euler angles and sometimes appear unexpected situations difficult to clarify.

Fig. 12.11 Examples of tire marks of the collision described in 12.1 and Figs. 12.2, 12.3, 12.4, 12.5, 12.6 and 12.7

4. In Chap. 13 we will make an introduction about a supremely important situation related with the point called **PNE** or *Point of No Escape*, because this concept has severe implications with the legal aspects that arise when a collision occurs.

12.3   Roadway Markings and Damaged Vehicle Inspection

They are many types of tire marks, that correspond to the marks left on the pavement by the vehicles involved in a collision, which can be very valuable for the traffic collision investigator, as physical evidence to determine the speeds and angles of the speed between other important information taken in the collision scene such as the witness statements.

Other traces of fundamental importance are those left by the vehicles in the collision itself, be they dents, scratching, paint stains, etc. (The paint stains that one vehicle can leave in the other are due to the intense heat that occurs in a collision that melts the paint).

The dents represent one of the most valuable sources in the research and reconstruction process as they are signs that provide to an experienced person, crucial indications of fundamental aspects of the dynamics of the collision, being able to determine for example which vehicle initiated the collision. In complex collisions, with roll over movement, the interpretation of damage to vehicles requires extreme care, as the collision does not occur on a plane, i.e. it is not two-dimensional but three-dimensional. In Chaps. 16 and 17 are given the expressions that allow to study these dynamic three-dimensional processes.

In Fig. 12.11 an example of tire marks left by the colliding vehicles described in 12.1 and Figs. 12.1, 12.2, 12.3, 12.4, 12.5, 12.6 and 12.7. Although the photos are not very clear, it can be seen the tire prints and the scratches on the asphalt.

In the Autostats database (www.4n6xprt.com), it is possible to find extremely useful information about the dimensions, masses, moments of inertia, location of the center of gravity, stability ratio, etc., of countless vehicles.

12.4   Safety Index or Vehicle Deformation

Different entities and researchers have tried to define a "severity index" of a collision, in order to be able to classify, sort and compare the damage that occurred in different collisions. We will mention by way of example, because of its historical importance one of the most used developed in the project Traffic Accident Data Index administered by the *National Safety Council* called the VDI "Vehicular Deformation Index" that considers: the direction of the main force at the point of impact, the location of the vehicle deformation, the general type of collision and a scale of damage, which we are going to clarify based on the excellent reference [8].

Description of the vehicular deformation index

The vehicular deformation index consists of *7 digits*, and includes *4 data*, which are briefly described below and illustrated through figures:

1. Direction of the main force at the point of impact
2. Location of vehicular deformation
3. General type of collision
4. Damage scale

DATA # 1: DIRECTION OF THE MAIN FORCE AT THE POINT OF IMPACT (Digits 1 and 2).

The direction of force is indicated with the help of a marked circle with the numbers 01 to 12, like a clock, which is placed at the point of impact, matching the longitudinal axis of the vehicle with line 06–12 (see Fig. 12.12). The first and the second digits are used for describing this data.

DATA # 2: LOCATION OF VEHICULAR DEFORMATION (Digits 3,4 and 5).

The third digit of the severity index generally describes the part of the vehicle most affected by the collision (see Fig. 12.13), while the fourth and fifth digits describe the

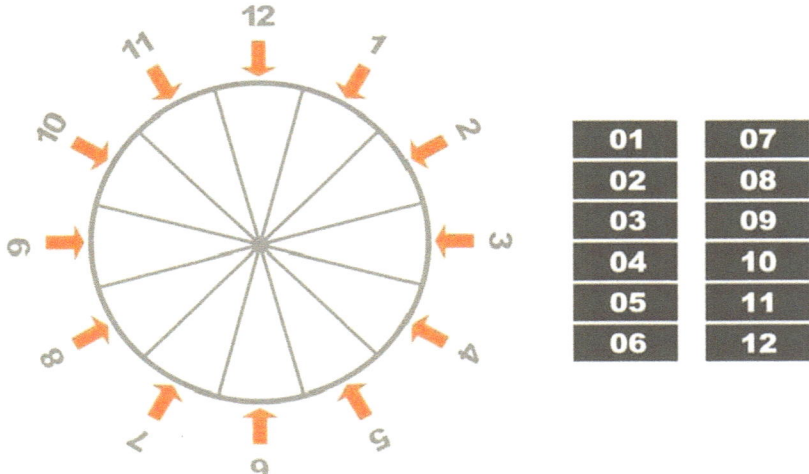

Fig. 12.12 Impact force direction

specific horizontal and vertical location of the damage respectively (Fig. 12.13). The damage in a vertical plane is indicated in the Fig. 12.13. The meaning of the letters is as follows:

For Fig. 12.13: Overview (Third Digit)

T: Up
U: Down
F: Front
B: Back
R: Right
L: Left

For Fig. 12.13: Damage in a horizontal plane (Fourth digit)

D: Distributed damage
L: Left or rear front damage
C: Front or rear center damage
R: Right or rear frontal damage
F;   Left side frontal damage
P: Damage to the right or left passenger area
B: Right or left side back damage
Y: Damage combination F y P o L y C.
Z: Damage combination B y P o R y C.

For Fig. 12.13: Damage in a vertical plane (Fifth digit)

A: Total vertical damage
H: Chassis damage to cap
E: All lower damage
G: All damage above the glass panels
M: Intermediate damage
L: Lower damage

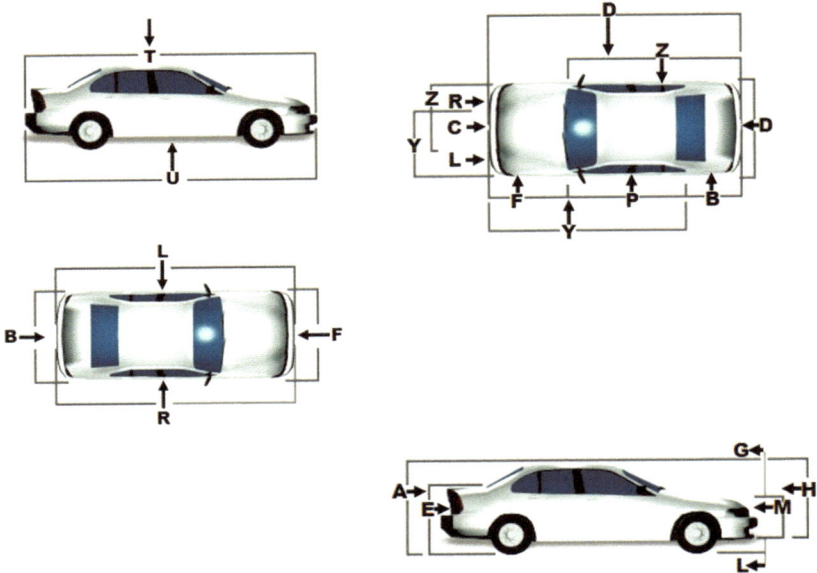

Fig. 12.13

DATA # 3: GENERAL TYPE OF COLLISION (Digit 6).

This digit indicates the general type of collision.:

W: Impact with a wide object (greater than 40.0 cm)
N: Impact with a narrow object (less than 40.0 cm)
S: Sidestroke
O: Wobble
F: Fire
Y: Fire with impact
Z: Immersion (when water represents a hazard to passengers)

DATA # 4: DAMAGE SCALE (Digit 7).

Finally, the digit # 7, describes the scale of damage from 1 to 9 (see Fig. 12.14)

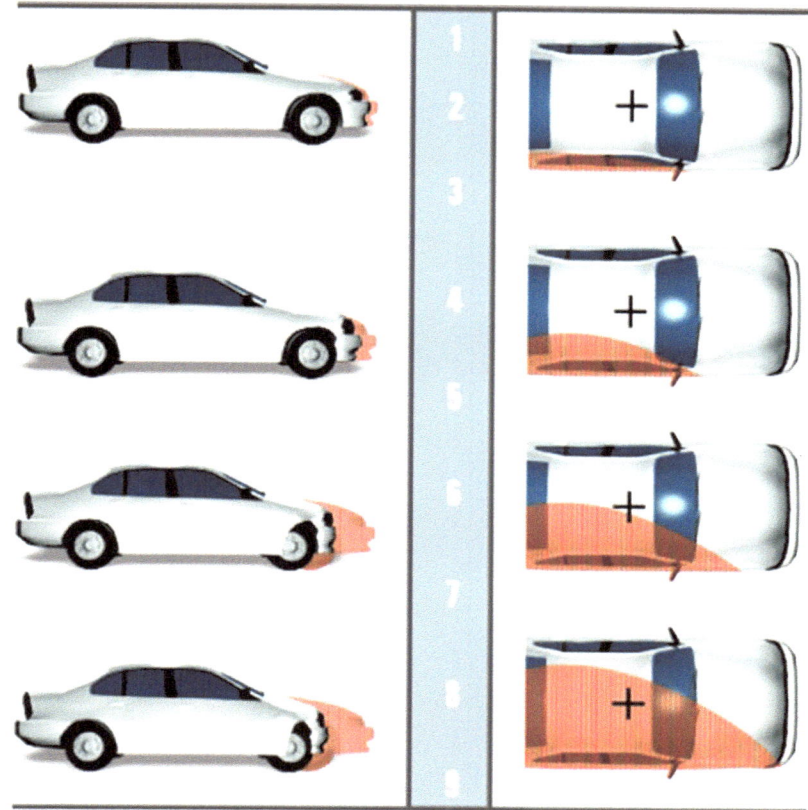

Fig. 12.14

Application Example

For a vehicle that had a lateral crash it would have to denote that the VDI is 06RDGW7 in which: the 06 refers to the direction in the circle system, R refers to the collision being on the right side, D and G refers to the specific horizontal and vertical location, W denotes that the object against which it collided were greater than 40.0 cm and 7 refers to the scale damage.

13.1 Determination of the Braking Visibility Distance

The braking visibility distance d_{VF} is the minimum distance necessary for a vehicle to stop before reaching a fixed obstacle that appears suddenly on the road, and it is the sum of the perception-reaction distance d_{PR} plus the braking distance d_F.

Perception-Reaction Distance

To calculate it is necessary to distinguish two phases in which the braking process takes place:

13.1.1 First Phase

The first phase is the one that takes place from the time the obstacle appears on the track until the driver applies the brakes and at the distance that the vehicle reaches in this phase is called *perception-reaction distance* as it occurs in the perception time and the reaction time, which are explained below:

- *The perception time* begin when the driver see the obstacle until to the moment that the decision is made about whether or not to brake and this time depends on the shape of the object, the visual conditions of the driver and the driver's willingness to brake.

Supplementary Information The online version contains supplementary material available at https://doi.org/10.1007/978-3-031-62355-4_13.

The perception time can be in the order of 0.5 s on urban roads and up to 2 s in rural areas, the latter being greater due to the lack of readiness of the driver to perform the braking maneuver.

- *The reaction time* is used by the driver to apply the brakes once he decides to do and depends on the reaction speed of the driver. This time is in the order of 0.5 to 1 s.

The sum of the time of perception and the reaction time is known as *psychotechnical time* t_{sic} and is in the order of 1.5 s on urban roads with the best visibility, that is, in optimal conditions. It is of the utmost importance to note, that the use of the cell phone while driving, increases this time by 1 s, which increases significantly, the risk of an accident. This fact reaffirms the urgent need for criminalizing those who drive their vehicle by talking on a cell phone.

In this way, the *perception-reaction distance*, d_{PR}, it can be easily calculated, being v, the speed at which the vehicle is traveling:

$$d_{PR} = v \times t_{sic} \tag{13.1}$$

13.1.2 Second Phase

The second phase that needs to be considered is the phase that runs from the time the driver applies the brakes until the vehicle stops completely and at the distance that the vehicle can travel in this time is called the *braking distance* d_F, which depends:

(a) On the friction factor of the tires against the pavement (see Table 13.1, based on ref 7) and,
(b) of the slope of the road

13.1.2.1 Breaking Distance

To find an expression that allows us to determine the braking distance, d_F, assume that the vehicle moves on a road that has an inclination of an angle β, as shown in Fig. 13.1 in which it is possible to identify three forces:

Table 13.1

Friction coefficients		
Rubber on concrete	Static	Kinetic
Dry pavement	1.00	0.80
Wet pavement	0.80	0.25

(a)

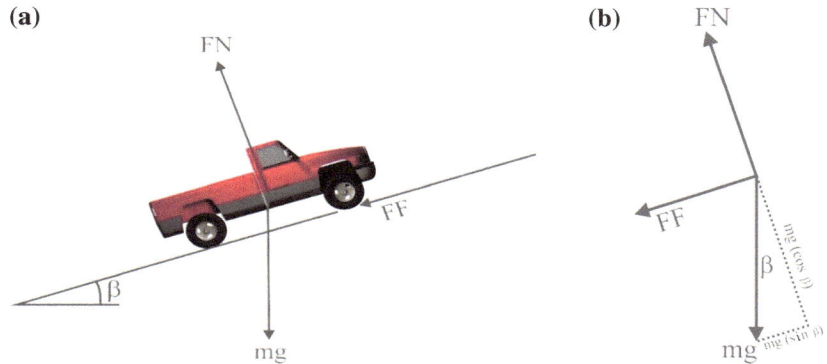

(b)

Fig. 13.1 **a** Forces acting on the vehicle. **b** Decomposition of forces on two perpendicular axles

- The frictional force of the wheels against the pavement, F_F
- The Normal force, F_N
- and the weight of the vehicle, mg

To facilitate the development of the problem, the Fig. 13.1 shows the weight components in the direction of movement and in the direction normal to the movement of the vehicle. As in the normal direction of movement there is no acceleration, taking a sum of forces in that direction we get:

$$F_N = mg\,(\cos\beta) \tag{13.2}$$

Now, taking a sum of forces in the direction of vehicle movement we obtain:

$$-F_F - mg\,(\sin\beta) = m\frac{dv}{dt}$$

As $F_F = \mu F_N$, where μ is the friction factor against the pavement, which normally has the values indicated in the Table 13.1:

In addition, as $\frac{dv}{dt} = \frac{dv}{dx} \times \frac{dx}{dt} = v\frac{dv}{dx}$ then:

$$-\mu F_N - mg\,(\sin\beta) = mv\frac{dv}{dx} \tag{13.3}$$

Replacing Eq. 13.2 in 13.3 we obtain:

$$-mg[\mu\cos\beta + \sin\beta]dx = mvdv$$

Now, integrating from the moment that the brakes are applied until the vehicle stops completely and the terms rearranged, we get the required expression for the braking distance d_F:

$$-\int_{0}^{d_F} g[\mu\cos(\beta) + \sin\beta]dx = \int_{v}^{0} vdv$$

$$-g[\mu\cos(\beta) + \sin\beta]d_F = -\frac{v^2}{2}$$

$$d_F = \frac{v^2}{2g[\mu\cos(\beta) + \sin\beta]}$$

$$d_F = \frac{v^2}{2g\cos\beta\left[\mu + \frac{\sin\beta}{\cos\beta}\right]}$$

$$d_F = \frac{v^2}{2g\cos\beta[\mu + \tan\beta]}$$

If P is the slope of the track and as $P = tan\ \beta$, then:

$$d_F = \frac{v^2}{2g\cos\beta[\mu + P]} \qquad (13.4)$$

Now, if we consider that slopes between –10 and 10%, the value of $cos\ \beta$ is between 0.995 and 1, that is $cos\ \beta \approx 1$, then it is possible to make the following approach, being negligible for the purposes of the study of vehicle collisions.

$$d_F = \frac{v^2}{2g[\mu + P]} \qquad (13.5)$$

So, the braking visibility distance, d_{VF} is obtained based on Eqs. 13.1 and 13.5:

$$d_{VF} = d_{PR} + d_F = v \times t_{sic} + \frac{v^2}{2g[\mu + P]} \qquad (13.6)$$

This equation can be easily programmed and is an extremely useful information, especially to determine aspects of civil and criminal liability.

13.2   Point of No Escape (PNE) and Applications. A Very Important Consideration in Crash Investigations

As an application of the calculation of the perception-reaction and the braking distance (Eq. 13.6), the situations indicated in Figs. 13.2, 13.3 and 13.4 will be explained. A new concept, the *Point of No Escape* will be introduced. Usual values for psychotechnical time of 1.5 s and a coefficient of friction of 0.8 for dry pavement are used.

Fig. 13.2 PNE from crash
point: 42.72 m (140ft)

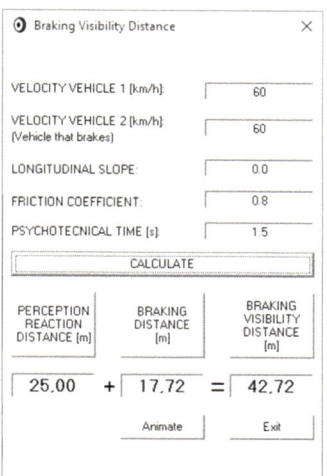

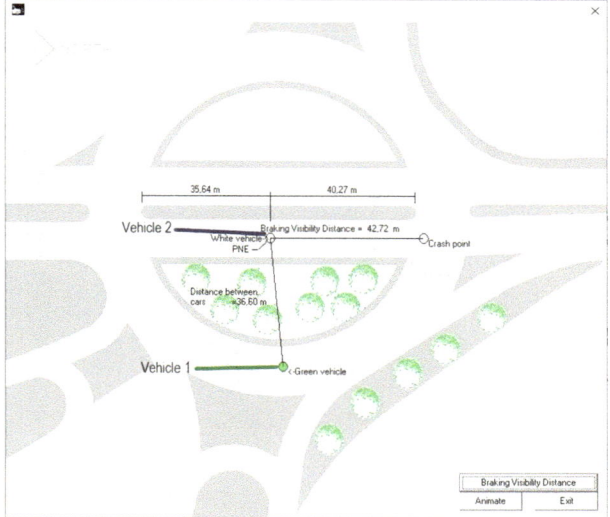

Initially we assume in Fig. 13.2 that vehicle 1 (green) travel through a roundabout at 60 km/h and has no stop priority at the intersection, while vehicle 2 (white) goes south-north on the straight track at a speed of 60 km/h, and, for some reason, vehicle 1 will not mark the mandatory stop at the intersection.

In this circumstances and in case that the two vehicles can see each other, the driver of vehicle 2, that have priority, to avoid the collision, should start the braking maneuver at 42.72 m from the crash point at the intersection, from a point called in the technical literature **PNE** (Point of No Escape), because if its distance (breaking visibility distance)

Fig. 13.3 PNE from crash
point: 64.82 m (213ft)

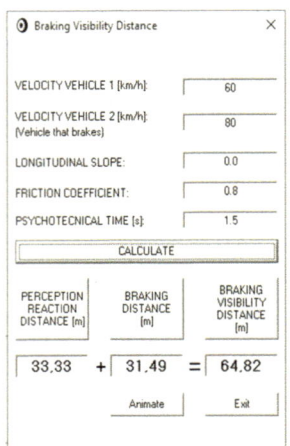

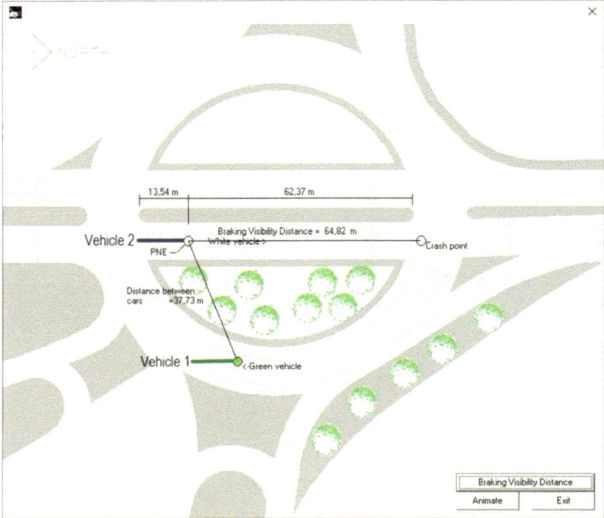

to the intersection was less, the driver of vehicle 2 could not carry out the tasks of perception, reaction and braking, to avoid the collision, accordingly Eq. 13.6 and the crash, will happen.

It should be noted, that the psychotechnical time assumed for average conditions of 1.5 s, should be increased if there are adverse visibility conditions or distracting conditions such as the use of the cell phone that can increase this time by 1 s. Similarly, if the pavement is wet, it increases the braking distance which can be twice that of a dry pavement. These two aspects, lead to the PNE being further away from the place where it must be braked to avoid a collision or to hit a pedestrian or cyclist.

Fig. 13.4 PNE from crash
point: 90.88 m (298ft)

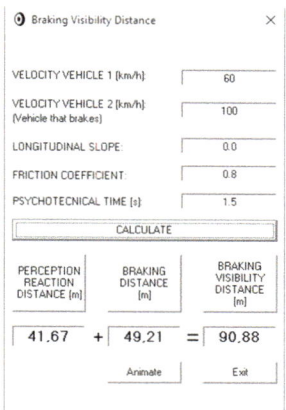

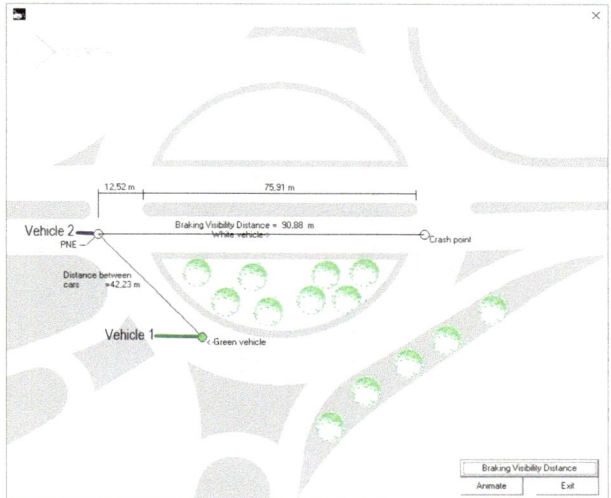

The Figs. 13.3 and 13.4 show the same situation, but considering that vehicle 2 would transit at two higher velocities, maintaining the other values:

(a) If the vehicle 2 were at 80 km/h the PNE would be at 64.82 m (Fig. 13.3) from intersection,

(b) and if it were at 100 km/h, the PNE would be at 90.88 m (Fig. 13.4) from the intersection respectively.

We can easily conclude that in these last two cases the collision would be inevitable.

The calculations and table shown in Figs. 13.2, 13.3 and 13.4 were programmed in Visual Basic and the programming of the equations of perception-reaction and braking times, allows to analyze in seconds, all situations that are required and therefore represent a valuable tool for the specialist and researcher of vehicular collisions.

In this chapter it is determined at which velocity a vehicle can move on a circular planar curve, banked or not, so that there is no sliding of the wheels on the pavement, or an overturn occurs.

Figure 14.1 shows a moving vehicle on a circular planar curve. The X axis points towards the center of the curve and the Y axis is tangential to it; so, for each position in which the vehicle is located, the perpendicular axes X and Y will be different.

Notation

m   Mass of the vehicle
v   Vehicle speed
R   Radius of the curve
μ Coefficient of friction between tires and pavement
y   Tangential axis to the curve
x   Axis with direction towards the center of the curve
F_F Frictional force of the wheels against the pavement
F_P Force normal to the surface
mg   Weight of the vehicle

L. G. Mejía Cañas, *Introduction to the Theory of Vehicular Collisions*, Synthesis Lectures on Mechanical Engineering, https://doi.org/10.1007/978-3-031-62355-4_14

Fig. 14.1 Vehicle moving on
a planar curve with radius R

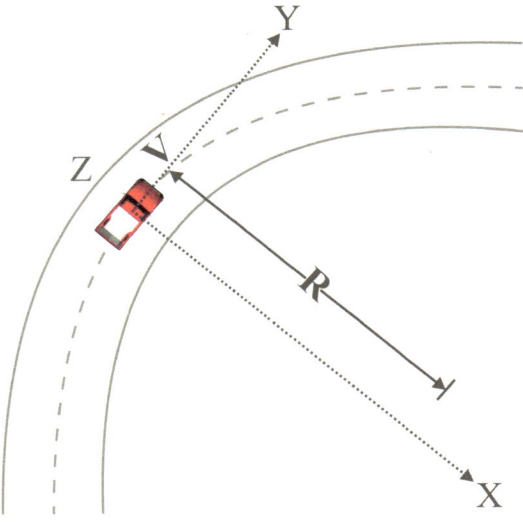

14.1   Unbanked Curves

The Fig. 14.2 show the free body diagram of the vehicle when the vehicle is on the curve. Three forces can be identified: F_P, F_F and mg

As in a vertical sense there is no acceleration, $F_P = mg$ and therefore $F_F = \mu(mg)$.

Now, as horizontally, only the frictional force is present, this is equal to the mass of the vehicle by the radial acceleration a R:

$$F_F = ma_R = m\frac{v^2}{R}$$

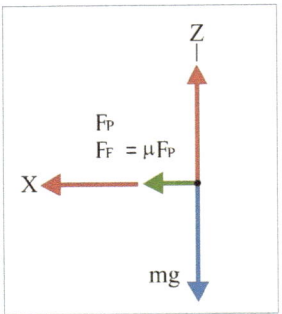

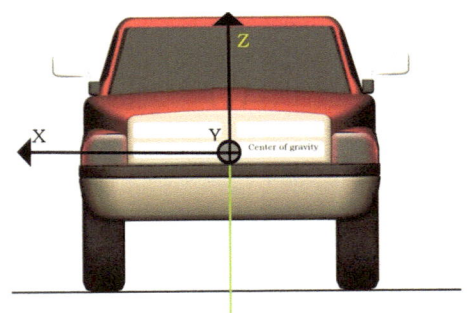

Fig. 14.2 Forces acting on a vehicle in an unbanked curve

So,

$$\mu(mg) = m\frac{v^2}{R}$$

And clearing v:

$$v = \sqrt{\mu g R} \tag{14.1}$$

With this equation, you can find the maximum speed at which the vehicle travels over the curve without slipping. If a vehicle takes a curve without a superelevation exceeding the speed v, not only is the maneuvering complicated by having difficulty controlling the car, but even, it can lose control, also increasing the perception-reaction time with respect to a situation on a straight track, being important to note that the previous equation does not depend on the mass of the vehicle.

14.2 Banked Curves

Suppose now that the road has a bank with an angle β. The forces present are the same as in the case of unbanked curves and likewise are the conclusions regarding the perception—reaction time and the loss of control of the vehicle, being able to present not only sliding but also overturning when the velocity v, for which was designed the bank is exceeded. In the design of roads the angle β is obtained for the so-called design speed.

As in a vertical direction there is no acceleration, taking sum of vertical forces we get:

$$F_P(\cos\beta) - mg - F_F(\sin\beta) = 0$$

And how $F_F = \mu F_P$ and cleaning for F_P we obtain:

$$F_P = \frac{mg}{\cos\beta - \mu\sin\beta}$$

Now, as in the horizontal direction the radial acceleration $a_R = m\frac{v^2}{R}$, then taking horizontal forces equilibrium:

$$F_P(\sin\beta) + F_F(\cos\beta) = ma_R = m\frac{v^2}{R}$$

And how $F_F = \mu F_P$ then:

$$F_P[\sin\beta + \mu(\cos\beta)] = m\frac{v^2}{R}$$

and replacing the value of F_P in the equation above.

$$\frac{mg}{\cos\beta - \mu\sin\beta}[\sin\beta + \mu(\cos\beta)] = m\frac{v^2}{R}$$

In this way, clearing v, we get:

$$v = \sqrt{\frac{gR[\sin\beta + \mu(\cos\beta)]}{\cos\beta - \mu\sin\beta}} \tag{14.2}$$

With this equation, it is possible to find the maximum speed at which the vehicle travels over the curve without slippage. It is important to note that the above equation, as well as in the case of unbanked curves, does not depend on the mass of the vehicle, and note that for $\beta = 0$ the Eq. 14.2 becomes Eq. 14.1.

From Agatha Christie's Poirot, to Reflect and Conclude:

The Mystery of the Blue Train

1. One fact is supported by another fact, and it may be that none of them is really a fact.
2. The least suspicious person can be the guiltiest.

The Silent Witness
There is only one world, the one that exists here and now, a physical accident with some biochemical ornaments.

In Chap. 7 was explained that vehicular collisions can be classified into two groups:

(a) Central collisions, which in turn are also divided into direct central collisions and oblique central collisions. It should be noted that the direct central collision is a particular case of an oblique central collision.
(b) Eccentric collisions.

The concept of the so-called "Crash Line" was also clarified in Chap. 7, and will be extremely useful for the approach of the equations that govern the dynamics of vehicular collisions of the types of collisions mentioned.

This chapter will address the approach of equations for *two-dimensional collisions*, that is, collisions in a plane. Later, the Chap. 17 will cover the subject of a collision in which vehicles rise from the ground, producing a *three-dimensional collision*, obviously much more complex than a two-dimensional collision, but required in some reconstruction studies.

L. G. Mejía Cañas, *Introduction to the Theory of Vehicular Collisions*, Synthesis Lectures on Mechanical Engineering, https://doi.org/10.1007/978-3-031-62355-4_15

15.1 Oblique Central Collision in Two Dimensions

For an oblique central collision, expressions will then be derived for the angle of the crash line, pre- and post-collision speeds and angles, considering the preservation of the amount of linear movement.

It is important to note that the deduction of the following equations will assume that the height of the center of mass of the colliding vehicles is approximately the same and therefore the analysis is two-dimensional.

However, when there is an appreciable difference between the heights of the mass centers, turns can occur around the vehicle axes x, and Y (Fig. 15.1) and the analysis becomes extremely complex.

In these cases, the equations deduced in Chap. 17, corresponding to the translational and rotational movement in three dimensions of a rigid body, and as already mentioned, should be applied for vehicles that rise from the floor during the collision.

Fig. 15.1 Oblique central collision: Position of vehicles at the time of collision. Note that the centers of mass of the vehicles (red circles) are located on a straight line, the crash line

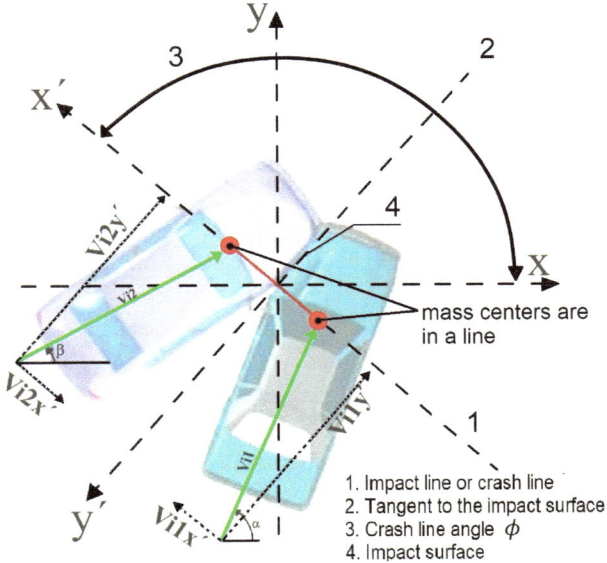

1. Impact line or crash line
2. Tangent to the impact surface
3. Crash line angle ϕ
4. Impact surface

Formulation of Basic Equations

Notation

$F\Delta t$:	Impulse that one vehicle exerts on the other at the moment of impact.
$V_{i1}, Vf1$:	Initial and final speeds of the center of mass of the vehicle 1.
V_{i2}, V_{f2}:	Initial and final speeds of the center of mass of the vehicle 2.
$V_{i1x'}, V_{i1y'}, V_{f1x'}, V_{f1y'}$:	Components in x', y' of the initial and final speeds of the center mass of vehicle 1.
$V_{i2x'}, V_{i2y'}, V_{f2x}, V_{f2y'}$:	Components in x', y' of the initial and final speeds of the center of mass of vehicle 2.
α, α':	Angles that form the initial and final speeds of the center of mass of vehicle 1 with the x axis.
β, β':	Angles that form the initial and final speeds of the center of mass of vehicle 2 with the x axis.
ϕ:	Crash line angle.

The axes *x*, *y* are arbitrary (they can match, for example, the north and south geodesic axes) and the axes *x/*, *y/* matches with the impact line (crash line) and the tangent to the impact surface, respectively (see Fig. 11).

To express $V_{i1x'}$ in terms of V_{i1} y $V_{i2x'}$ in terms of V_{i2} it is necessary to keep in mind that the angles α and β are referenced to the axis *x*, therefore, with regard to the crash line (axis *x/*), the corresponding angles will be $(\alpha - \varphi)$ y $(\beta - \varphi)$.

In this way,

$$V_{i1x'} = V_{i1}cos(a - \phi)$$

$$V_{i2x'} = V_{i2}cos(\beta - \phi)$$

It also happens with $V_{f1x'}$ y $V_{f2x'}$:

$$V_{f1x'} = V_{f1}cos(\alpha - \phi)$$

$$V_{f2x'} = V_{f2}cos(\beta - \phi)$$

(a) Movement parallel to the crash line (Direction X'):

According to Newton's third law, impulses $F\Delta t$ have equal magnitude and opposite direction, therefore, the forces cancel each other and then the theorem of the conservation of the amount of linear motion can be applied. Assuming that there is no loss of mass:

$$m_1 V_{i1x'} + m_2 V_{i2x'} = m_1 V_{f1x'} + m_2 V_{f2x'} \tag{15.1}$$

Applying the concept of the coefficient of restitution discussed in the Chap. 8, we obtain:

$$V_{f2x'} - V_{i1x'} = e(V_{i1x'} - V_{i2x'}) \tag{15.2}$$

(b) Movement perpendicular to the crash line (direction Y')

Because the momentum at the time of the collision occurs only in the direction x', the respective components of the speeds in the direction y' are preserved after the crash. For this reason, it can be stated:

$$V_{i1y'} = V_{f1y'} \tag{15.3}$$

$$V_{i2y'} = V_{f2y'} \tag{15.4}$$

15.1.1 Angle of the Crash Line ϕ in Function of the Initial and Final Speeds

Assuming that the initial and final speeds and angles are known, it is possible to find the crash line angle as follows:

Replacing the starting and final speeds, for example, of vehicle 2, Eq. 15.4 provides:

$$V_{i2}\sin(\beta - \phi) = V_{f2}\sin(\beta' - \phi)$$

$$V_{i2}(\sin\beta\cos\phi - \sin\phi\cos\beta) = V_{f2}\left(\sin\beta'^{\cos\phi - \sin\phi\cos\beta'}\right)$$

$$V_{i2}\sin\beta\cos\phi - V_{i2}\sin\phi\cos\beta = V_{f2}\sin\beta'^{\cos\phi - V_{f2}\sin\phi\cos\beta'}$$

$$\left(V_{f2}\cos\beta' - V_{i2}\cos\beta\right)\sin\phi = \left(V_{f2}\sin\beta' - V_{i2}\sin\beta\right)\cos\phi$$

$$\frac{\sin\phi}{\cos\phi} = \frac{V_{f2}\sin\beta' - V_{i2}\sin\beta}{V_{f2}\cos\beta' - V_{i2}\cos\beta}$$

$$\tan\phi = \frac{V_{f2}\sin\beta' - V_{i2}\sin\beta}{V_{f2}\cos\beta' - V_{i2}\cos\beta}$$

$$\phi = \tan^{-1}\left[\frac{V_{f2}\sin\beta' - V_{i2}\sin\beta}{V_{f2}\cos\beta' - V_{i2}\cos\beta}\right] \tag{15.5}$$

Similarly, the crash line angle can be found with the components in y' of the initial and final speeds of vehicle 1, obtaining:

$$\phi = \tan^{-1}\left[\frac{V_{f1}\sin\alpha' - V_{i1}\sin\alpha}{V_{f1}\cos\alpha' - V_{i2}\cos\alpha}\right] \tag{15.6}$$

The value obtained for the crash line angle (ϕ) with Eq. 15.5, it is the same as that obtained with Eq. 15.6.

15.1.2 Final Conditions as a Function of: The Initial Conditions, the Crash Line Angle and the Restitution Coeficient

Assuming that the initial conditions (speeds and angles), the crash line angle, and the restitution coefficient are known, it is possible to find the final conditions:
Solving for $V_{f2x'}$ from Eq. 15.2:

$$V_{f2x'} = eV_{i1x'} - eV_{i2x'} + V_{f1x'} \tag{15.7}$$

Replacing Eq. 15.7 in Eq. 15.1, we have:

$$m_1 V_{i1x'} + m_2 V_{i2x'} = m_1 V_{f1x'} + m_2 e V_{i1x'} - m_2 e V_{i2x'} + m_2 V_{f1x'}$$

$$V_{f1x'}(m_1 + m_2) = V_{i1x'}(m_1 - m_2 e) + V_{i2x'}(m_2 + m_2 e)$$

$$V_{f1x'} = \frac{V_{i1x'}(m_1 - m_2 e) + V_{i2x'}(m_2 + m_2 e)}{m_1 + m_2}$$

$$V_{f1x'} = \frac{V_{i1}\cos(\alpha - \phi)(m_1 - m_2 e) + V_{i2}\cos(\beta - \phi)(m_2 + m_2 e)}{m_1 + m_2} \tag{15.8}$$

By replacing Eq. 15.8 in Eq. 15.7, we obtain:

$$V_{f2x'} = eV_{i1x'} - eV_{i2x'} + \frac{V_{i1x'}(m_1 - m_2 e) + V_{i2x'}(m_2 + m_2 e)}{m_1 + m_2}$$

$$V_{f2x'} =$$

$$\frac{V_{i1x'}em_1 + V_{i1x'}em_2 - V_{i2x'}em_1 - V_{i2x'}em_2 + V_{i1x'}m_1 - V_{i1x'}em_2 + V_{i2x'}m_2 + V_{i2x'}em_2}{m_1 + m_2}$$

$$V_{f2x'} = \frac{V_{i1x'}(m_1 - m_1 e) + V_{i2x'}(m_2 - m_1 e)}{m_1 + m_2}$$

$$V_{f2x'} = \frac{V_{i1}\cos(\alpha - \phi)(m_1 + m_1 e) + V_{i2}\cos(\beta - \phi)(m_2 - m_1 e)}{m_1 + m_2} \tag{15.9}$$

From Eq. 15.3:

$$V_{f1y'} = V_{i1y'}$$

$$V_{f1y'} = V_{i1}\sin(\alpha - \phi) \tag{15.10}$$

And from Eq. 15.4:

$$V_{f2y'} = V_{i2y'}$$

$$V_{f2y'} = V_{i2}\sin(\beta - \phi) \tag{15.11}$$

(a) Therefore, the final speed of *vehicle 1* will be:

$$V_{f1} = \sqrt{\left[\frac{V_{i1}\cos(\alpha - \phi)(m_1 - m_2 e) + V_{i2}\cos(\beta - \phi)(m_2 + m_2 e)}{m_1 + m_2}\right]^2 + [V_{i1}\sin(\alpha - \phi)]^2} \tag{15.12}$$

- its final angle with respect to the x' axis:

$$(\alpha' - \phi) = \tan^{-1}\frac{V_{i1}\sin(\alpha - \phi)(m_1 + m_2)}{V_{i1}\cos(\alpha - \phi)(m_1 - m_2 e) + V_{i2}\cos(\beta - \phi)(m_2 + m_2 e)}$$

- and its final angle relative to the axis x:

$$\alpha' = \phi + \tan^{-1} \frac{V_{i1}\sin(\alpha - \phi)(m_1 + m_2)}{V_{i1}\cos(\alpha - \phi)(m_1 - m_2e) + V_{i2}\cos(\beta - \phi)(m_2 + m_2e)} \quad (15.13)$$

(b) Likewise, the final speed of <u>vehicle 2</u> will be:

$$V_{f2} = \sqrt{\left[\frac{V_{i1}\cos(\alpha - \phi)(m_1 + m_1e) + V_{i2}\cos(\beta - \phi)(m_2 - m_1e)}{m_1 + m_2}\right]^2 + [V_{i2}\sin(\beta - \phi)]^2}$$

$$(15.14)$$

- its final angle with respect to the x' axis

$$(\beta' - \phi) = \tan^{-1} \frac{V_{i2}\sin(\beta - \phi)(m_1 + m_2)}{V_{i1}\cos(\alpha - \phi)(m_1 + m_1e) + V_{i2}\cos(\beta - \phi)(m_2 - m_1e)}$$

- and its final angle relative to the axis x:

$$\beta' = \phi + \tan^{-1} \frac{V_{i2}\sin(\beta - \phi)(m_1 + m_2)}{V_{i1}\cos(\alpha - \phi)(m_1 + m_1e) + V_{i2}\cos(\beta - \phi)(m_2 - m_1e)} \quad (15.15)$$

15.1.3 Initial Conditions as a Function of: The Final Conditions, the Crash Line Angle and the Restitution Coefficient

Assuming that the final conditions (speeds and angles), the angle of the bump line, and the restitution coefficient are known, it is possible to find the initial conditions:

Clearing $V_{i2x'}$ of the Eq. 15.2:

$$V_{i2x'} = V_{i1x'} - \frac{V_{f2x'} - V_{f1x'}}{e} \quad (15.16)$$

Replacing Eq. 15.3.1 in Eq. 15.1:

$$m_1 V_{i1x'} + m_2 V_{i1x'} - \frac{m_2 V_{f2x'}}{e} + \frac{m_2 V_{f1x'}}{e} = m_1 V_{f1x'} + m_2 V_{f2x'}$$

$$V_{i1x'}(m_1 + m_2) = m_1 V_{f1x'} + m_2 V_{f2x'} + \frac{m_2 V_{f2x'}}{e} - \frac{m_2 V_{f1x'}}{e}$$

$$V_{i1x'} = \frac{V_{f1x'}\left(m_1 - \frac{m_2}{e}\right) + V_{f2x'}\left(m_2 + \frac{m_2}{e}\right)}{(m_1 + m_2)}$$

$$V_{i1x'} = \frac{V_{f1}\cos(\alpha' - \phi)\left(m_1 - \frac{m_2}{e}\right) + V_{f2}\cos(\beta' - \phi)\left(m_2 + \frac{m_2}{e}\right)}{(m_1 + m_2)} \tag{15.17}$$

By replacing Eq. 15.17 in Eq. 15.16, we obtain:

$$V_{i2x'} = \frac{V_{f1x'}\left(m_1 - \frac{m_2}{e}\right) + V_{f2x'}\left(m_2 + \frac{m_2}{e}\right)}{(m_1 + m_2)} - \frac{V_{f2x'} - V_{f1x'}}{e}$$

$$V_{i2x'} = \frac{V_{f1x'}\left(m_1 - \frac{m_2}{e}\right) + V_{f2x'}\left(m_2 + \frac{m_2}{e}\right)}{(m_1 + m_2)} - \frac{\left(V_{f2x'}\frac{m_1}{e} + V_{f2x'}\frac{m_2}{e} - V_{f1x'}\frac{m_1}{e} - V_{f1x'}\frac{m_2}{e}\right)}{(m_1 + m_2)}$$

$$V_{i2x'} = \frac{V_{f1x'}\left(m_1 - \frac{m_1}{e}\right) + V_{f2x'}\left(m_2 - \frac{m_1}{e}\right)}{(m_1 + m_2)}$$

$$V_{i2} = \frac{V_{f1}\cos(\alpha' - \phi)\left(m_1 + \frac{m_1}{e}\right) + V_{f2}\cos(\beta' - \phi)\left(m_2 - \frac{m_1}{e}\right)}{(m_1 + m_2)} \tag{15.18}$$

From Eq. 15.3:

$$V_{i1y'} = V_{f1y'}$$

$$V_{i1y'} = V_{f1}\sin(\alpha' - \phi) \tag{15.19}$$

And from Eq. 15.4:

$$V_{i2y'} = V_{f2y'}$$

$$V_{i2y'} = V_{f2}\sin(\beta' - \phi) \tag{15.20}$$

(a) Therefore, the initial speed of *vehicle 1* will be:

$$V_{i1} = \sqrt{\left[\frac{V_{f1}\cos(\alpha' - \phi)\left(m_1 - \frac{m_2}{e}\right) + V_{f2}\cos(\beta' - \phi)\left(m_2 + \frac{m_2}{e}\right)}{(m_1 + m_2)}\right]^2 + \left[V_{f1}\sin(\alpha' - \phi)\right]^2} \tag{15.21}$$

- its initial angle with respect to the x' axis:

$$(\alpha - \phi) = \tan^{-1} \frac{V_{f1}\sin(\alpha' - \phi)(m_1 + m_2)}{V_{f1}\cos(\alpha' - \phi)\left(m_1 - \frac{m_2}{e}\right) + V_{f2}\cos(\beta' - \phi)\left(m_2 + \frac{m_2}{e}\right)}$$

- and its initial angle relative to the x-axis:

$$\alpha = \phi + \tan^{-1} \frac{V_{f1}\sin(\alpha' - \phi)(m_1 + m_2)}{V_{f1}\cos(\alpha' - \phi)\left(m_1 - \frac{m_2}{e}\right) + V_{f2}\cos(\beta' - \phi)\left(m_2 + \frac{m_2}{e}\right)} \tag{15.22}$$

(b) Likewise, the initial speed of *vehicle 2* will be:

$$V_{i2} = \sqrt{\left[\frac{V_{f1}\cos(\alpha' - \phi)\left(m_1 + \frac{m_1}{e}\right) + V_{f2}\cos(\beta' - \phi)\left(m_2 - \frac{m_1}{e}\right)}{(m_1 + m_2)}\right]^2 + \left[V_{f2}\sin(\beta' - \phi)\right]^2} \tag{15.23}$$

(c) its initial angle with respect to the x' axis:

$$(\beta - \phi) = \tan^{-1} \frac{V_{f2}\sin(\beta' - \phi)(m_1 + m_2)}{V_{f1}\cos(\alpha' - \phi)\left(m_1 + \frac{m_1}{e}\right) + V_{f2}\cos(\beta' - \phi)\left(m_2 - \frac{m_1}{e}\right)}$$

(d) and its initial angle with respect to the axis x:

$$\beta = \phi + \tan^{-1} \frac{V_{f2}\sin(\beta' - \phi)(m_1 + m_2)}{V_{f1}\cos(\alpha' - \phi)\left(m_1 + \frac{m_1}{e}\right) + V_{f2}\cos(\beta' - \phi)\left(m_2 - \frac{m_1}{e}\right)} \tag{15.24}$$

15.2   Eccentric Collision in Two Dimensions

In an eccentric collision the center of mass of colliding vehicles moves in a straight line (see Fig. 5), and therefore it is possible to analyze its displacement by applying the conservation of the amount of linear motion.

The equations that describe this displacement are derived below, recalling that in such collisions, the crash line does not join the centers of mass of the vehicles as shown in Fig. 15.2 (see also numeral 7.2). The conventions are the same as those shown in Fig. 15.1 with the additional terms due to rotation.

The equations to describe this type of collision are more complex than those already presented for an oblique central collision and are derived to have complete information about these two-dimensional equations. However at the end of this chapter, in number 15.2.4 a very important conclusion is reached for practical calculations assuming that they are carried out with judgment and common sense.

Fig. 15.2 Eccentric collision: Position of vehicles at the time of collision. Note that the centers of mass of the vehicles (red circles) are not located on a straight line

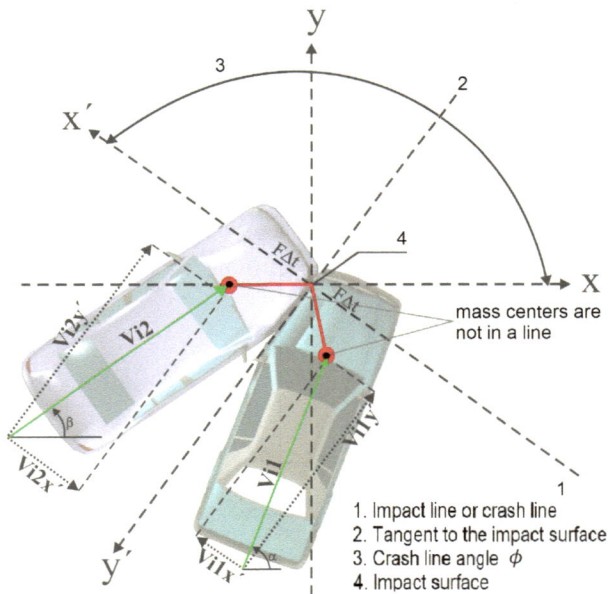

1. Impact line or crash line
2. Tangent to the impact surface
3. Crash line angle ϕ
4. Impact surface

15.2.1 Formulation of Basic Equations

Notation

P:	Point of contact between vehicles at the time of collision.
$F\Delta t$:	Impulse that one vehicle exerts on the other at the moment of impact.
V_{i1}, $Vf1$:	Initial and final speeds of the center of mass of the vehicle 1.
V_{i2}, V_{f2}:	Initial and final speeds of the center of mass of the vehicle 2.
$V_{i1x'}$, $V_{i1y'}$, $V_{f1x'}$, $V_{f1y'}$:	Components in x', y' of the initial and final speeds of the center mass of vehicle 1.
$V_{i2x'}$, $V_{i2y'}$, V_{f2x}, $V_{f2y'}$:	Components in x', y' of the initial and final speeds of the center of mass of vehicle 2.
V_{Pi1}, V_{Pf1}:	Initial and final speeds of contact point P 1 $_{Pi2}$.
V, V_{Pf2}:	Initial and final speeds of contact point P 2.
$V_{Pi1x'}$, $V_{Pi1y'}$, $V_{Pf1x'}$, $V_{Pf1y'}$:	Components in x', y' of the initial and final speeds of the point of Contact P 1.
$V_{Pi2x'}$, $V_{Pi2y'}$, $V_{Pf2x'}$, $V_{Pf2y'}$:	Components in x', y' of the initial and end speeds of the point of Contact P 2.
ω_{f1}, $\omega f2$	Final angular speeds of vehicles 1 and 2.
$\overrightarrow{r_{P1}}, \overrightarrow{r_{P2}}$:	Vectors from the center of mass to contact point P for vehicles 1 and 2.
α, α':	Angles that form the initial and final speeds of the center of mass of vehicle 1 with the x axis.
β, β':	Angles that form the initial and final speeds of the center of mass of vehicle 2 with the x axis.
ϕ:	Crash line angle.

- Movement Parallel to the Crash Line (Direction x´):

Likewise, as in oblique central collisions, in the direction parallel to the crash line, applying the conservation of the amount of linear motion, we get:

$$m_1 V_{i1x'} + m_2 V_{i2x'} = m_1 V_{f1x'} + m_2 V_{f2x'} \qquad (15.25)$$

However, in order to apply the concept of the coefficient of restitution explained in Chap. 8, it is necessary to consider that the relative speeds of separation and approximation refer, in this case, to the point of contact between the vehicles P:

$$e = \frac{V_{Pf2x'} - V_{Pf1x'}}{V_{Pi1x'} - V_{Pi2x'}}$$

Assuming that cars don't spin before the collision, then ($V_{Pi1x'} = V_{i1x'}$; $V_{Pi2x'} = V_{i2x'}$)
And we finally obtain:

$$V_{Pf2x'} - V_{Pf1x'} = e(V_{i1x'} - V_{i2x'}) \tag{15.26}$$

- Movement perpendicular to the crash line (direction y'):

In this direction, the velocity components are conserved after the collision:

$$V_{i1y'} = V_{f1y'}$$

$$V_{i2y'} = V_{f2y'}$$

15.2.2 Final Speed of Point P (Point of Contact)

a. For *vehicle 1:*

$$\overrightarrow{V_{Pf1}} = \overrightarrow{V_{f1}} + \overrightarrow{\omega_{f1}} \times \overrightarrow{r_{P1}}$$

$$(V_{Pf1x'})\hat{i} + (V_{Pf1y'})\hat{j} = (V_{f1x'})\hat{i} + (V_{f1y'})\hat{j} + \begin{vmatrix} \hat{i} & \hat{j} & \hat{k} \\ 0 & 0 & \omega_{f1} \\ r_{P1x'} & r_{P1y'} & 0 \end{vmatrix}$$

where $\hat{i}$, $\hat{j}$ are unit vectors in directions x', y' respectively.
   In this way:

$$(V_{Pf1x'})\hat{i} + (V_{Pf1y'})\hat{j} = (V_{f1x'})\hat{i} + (V_{f1y'})\hat{j} - (r_{P1y'}\omega_{f1})\hat{i} + (r_{P1x'}\omega_{f1})\hat{j}$$

And expressing the above equation by components is obtained:

$$V_{Pf1x'} = V_{f1x'} - r_{P1y'}\omega_{f1} \tag{15.27}$$

$$V_{Pf1y'} = V_{f1y'} + r_{P1x'}\omega_{f1} \tag{15.28}$$

b. For *Vehicle 2*:

The procedure is analogous that for the case of vehicle 1, obtaining:

$$V_{Pf2x'} = V_{f2x'} - r_{P2y'}\omega_{f2} \tag{15.29}$$

$$V_{Pf2y'} = V_{f2y'} + r_{P2x'}\omega_{f2} \tag{15.30}$$

c. Motion perpendicular to the crash line:

In this direction, the speed components are conserved after the crash:

$$V_{i2y'} = V_{f1y'}$$

$$V_{i2y'} = V_{f2y'}$$

15.2.3   Conservation of Angular Momentum with Respect to Point P

a. *For Vehicle 1:*

$$\left(\overrightarrow{r_{P1}} \times m_1 \overrightarrow{V_{i1}}\right)\hat{k} + I_1\omega_{i1} = \left(\overrightarrow{r_{P1}} \times m_1 \overrightarrow{V_{f1}}\right)\hat{k} + I_1\omega_{f1}$$

where I_1 it's the moment of inertia of vehicle 1 with respect to a vertical axis that passes through the point P. And, since it's supposed that cars don't have initial angular velocity ($\omega_{i1} = 0$):

$$\left(\overrightarrow{r_{P1}} \times m_1 \overrightarrow{V_{i1}}\right)\hat{k} = \left(\overrightarrow{r_{P1}} \times m_1 \overrightarrow{V_{f1}}\right)\hat{k} + I_1\omega_{f1}$$

$$m_1 \begin{vmatrix} \hat{i} & \hat{j} & \hat{k} \\ r_{P1x'} & r_{P1y'} & 0 \\ V_{i1x'} & V_{i1y'} & 0 \end{vmatrix}\hat{k} = m_1 \begin{vmatrix} \hat{i} & \hat{j} & \hat{k} \\ r_{P1x'} & r_{P1y'} & 0 \\ V_{f1x'} & V_{f1y'} & 0 \end{vmatrix}\hat{k} + I_1\omega_{f1}$$

$$m_1\left(r_{P1x'}V_{i1y'} - r_{P1y'}V_{i1x'}\right)\hat{k}\cdot\hat{k} = m_1\left(r_{P1x'}V_{f1y'} - r_{P1y'}V_{f1x'}\right)\hat{k}\cdot\hat{k} + I_1\omega_{f1}$$

$$m_1 r_{P1x'}V_{i1y'} - m_1 r_{P1y'}V_{i1x'} = m_1 r_{P1x'}V_{f1y'} - m_1 r_{P1y'}V_{f1x'} + I_1\omega_{f1} \tag{15.31}$$

b. *For Vehicle 2:*

The procedure is analogous that for the case of car 1, obtaining:

$$m_2 r_{P2x'}V_{i2y'} - m_2 r_{P2y'}V_{i2x'} = m_2 r_{P2x'}V_{f2y'} - m_2 r_{P2y'}V_{f2x'} + I_2\omega_{f2} \tag{15.32}$$

15.2.4 Conclusion

As can be seen in Eqs. 15.21 to 15.32, if pre-collision conditions are considered known, the unknowns would be:

$$V_{f1x'}; V_{f2x'}; V_{Pf1x'}; V_{Pf2x'}; V_{Pf1y'}; V_{Pf2y'}; \omega_{f1}; \omega_{f2}$$

That is, 8 equations with 8 unknowns would be made up, forming an 8×8 system of linear equations, which is not easy to solve manually.

In addition, there is the additional difficulty of being necessary to know exactly the point of contact between the vehicles, located vectoral with respect to the axles x', y'.

In the light of the above, it is reasonable to simplify the problem by *not* considering the rotation and applying the expressions presented in this chapter, since the errors obtained are acceptable for a rigorous but approximate collision analysis.

Historical Note: **Denis Poisson (1781–1840) [2].**

His work was based on the research of Wallis, Wren and Huygens related to the laws of the impulse and energy. In his book *Traité de mécanique* he dealt mainly with the collision of bodies with different shapes.

Sometimes, the description of the possible turns of a car in the post-collision phase require the consideration of three dimensions for the development of motion equations, as will be seen in next Chapter. This situation arises when, for some reason, one or both vehicles are lifted from the plane, a situation in which two-dimensional analyses can be used as an initial approximation, but, depending on the severity of the collision, sometimes three-dimensional dynamics must be used, in which it is necessary to define the position of the vehicle at any instant, which is achieved with the help of the so-called Euler angles. Initially it is necessary to know which are the types of turns, that should be considered in an three-dimensional crash investigation, i.e.: 1. Rollover, 2. Flip and 3. Spin which we will explain briefly in this chapter. Related with this matter, the excellent content of reference 1 is highly recommended for those who want to delve deeper into this topic.

16.1 "Rollover"

A "Rollover" occurs when, by some circumstance, the vehicle rotates around its longitudinal axis (Fig. 16.1). During this turn, four instability situations can occur, that we call turns of $90°$, 180^0, 270^0 and 360^0, and which we will analyze below.

a. Turn of $90°$

For a $90°$ turn and that the vehicle is on the wheels of one side, it is necessary that, as long as the vehicle rotate on those wheels, its center of mass rises and exceeds a height $h1$ (Fig. 16.3) which is equal to:

L. G. Mejía Cañas, *Introduction to the Theory of Vehicular Collisions*, Synthesis Lectures on Mechanical Engineering, https://doi.org/10.1007/978-3-031-62355-4_16

Fig. 16.1 Rollover

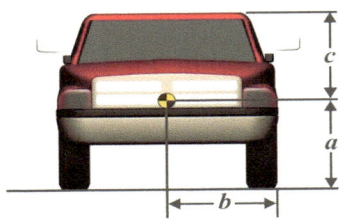 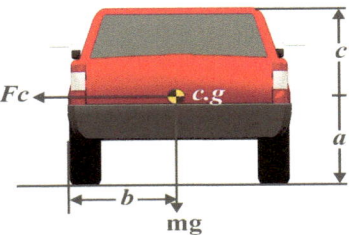

Fig. 16.2 Position of the center of mass of the vehicle and definition of values "a", "b" and "c". Front and rear view

$$h_1 = \sqrt{a^2 + b^2} \tag{16.1}$$

As the tires come off the road and the car begins to rotate, the height of the center of mass begins to increase, the same thing that does the vehicle's local potential energy. In this way, for there to be a turn of 90° it is necessary that the conversion of kinetic energy into work be enough to lift the center of mass above a height **h1**. If this happens, the vehicle will continue to rotate and the height of the center of mass will begin to decrease to a value **b**, for which the vehicle lies on its side.

So, **h1** is the height of the center of mass for which the vehicle reaches a point of maximum potential energy and a turn of 90° occurs. In addition, it is equivalent to a rotation of an angle equal to:

$$a_1 = \tan^{-1}\left(\frac{b}{a}\right) \tag{16.2}$$

Fig. 16.3 Maximum potential energy for a 90° turn

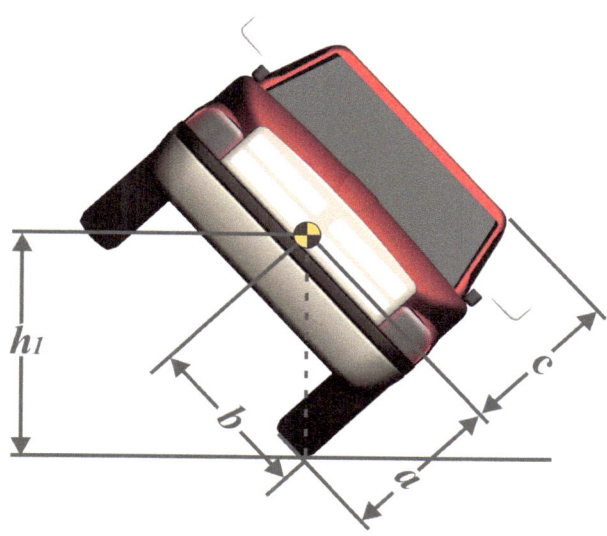

b. Turn of 180°

If, after a 90-degreel turn, there is still enough energy left to keep the car rotating, it may be on its roof completing a turn of 180°.

For this to happen, the height of the center of mass needs to start increasing again from **b** and reach a height **h2** (Fig. 16.4) given by the following expression:

$$h_2 = \sqrt{b^2 + c^2} \qquad (16.3)$$

Thus, the total angle of rotation that the vehicle must exceed to have a turn of 180° is:

$$a_2 = 90° + \tan^{-1}\left(\frac{c}{b}\right) \qquad (16.4)$$

So, **h2** it is then a second maximum of potential energy, and when this height is exceeded by the center of mass, it begins to descend until it is at a height **c** from the floor (the vehicle is on its roof).

c. Turn of 270°

For a 270° turn to occur, it is necessary that the height of the vehicle's center of mass begins to increase from **c** to a third potential maximum **h3** (Fig. 16.5) and exceeds it. The value of **h3** can be calculated using the following expression:

$$h_3 = \sqrt{b^2 + c^2} \qquad (16.5)$$

Fig. 16.4 Maximum potential energy for a 180° turn

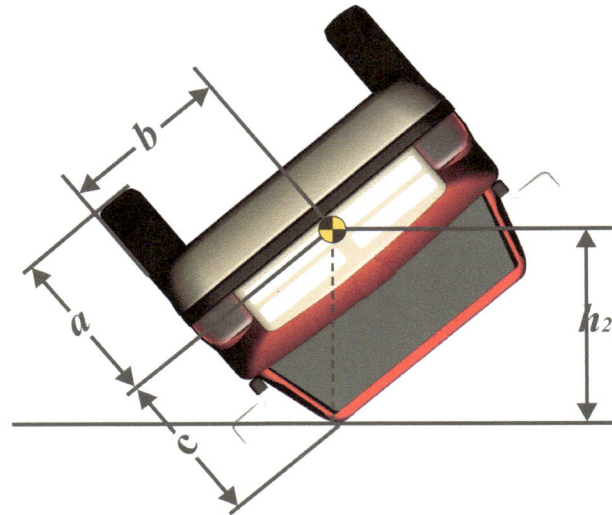

And the total angle of rotation that the vehicle must overcome for it to turn 270° is:

$$a_3 = 180° + \tan^{-1}\left(\frac{c}{b}\right) \tag{16.6}$$

After the third potential energy maximum has been exceeded, the height of the center of mass begins to decrease until it is at a height **b** from the floor (the vehicle lies on its other side).

Fig. 16.5 Maximum potential energy for a 270° turn

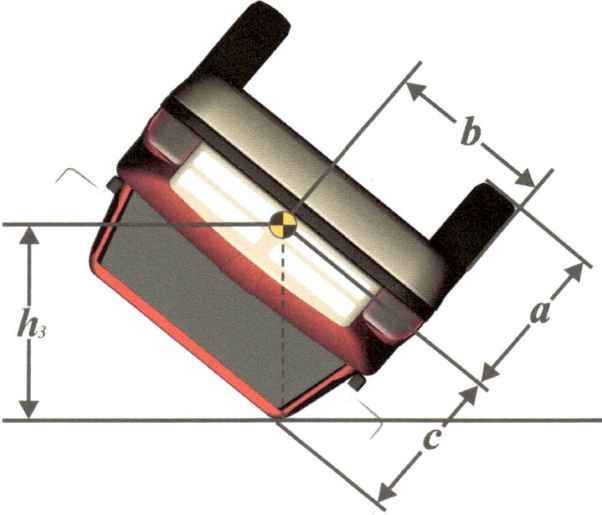

d. Turn of 360°

For a bell turn to occur 360°, it is necessary that the height of the center of mass of the vehicle begins to increase from b to a maximum quarter potential $h4$ (see Fig. 16.6) and get over it. The value of $h4$ can be calculated using the following expression:

$$h_4 = \sqrt{a^2 + b^2} \qquad (16.7)$$

The total angle of rotation that the vehicle must exceed in order to turn 360° is:

$$a_4 = 270° + \tan^{-1}\left(\frac{b}{a}\right) \qquad (16.8)$$

After exceeding the maximum quarter potential energy, the height of the center of mass begins to decrease to a height a from the floor (the vehicle returns to its initial position).

a. Graphical Representation of the Types of Turns

Derived from Fig. 16.3 [1] , the Fig. 16.7 graphically shows the information discussed in paragraphs "a" to "d".
Annex: Work Performed by a Vehicle that turns [Adapted from reference 1]
The excellent content of reference 1 is highly recommended for those who want to delve deeper into this topic.

Fig. 16.6 Maximum potential energy for a 360^0 turn

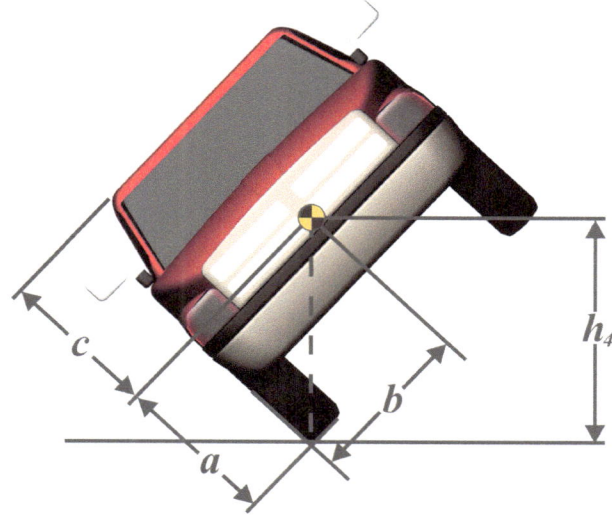

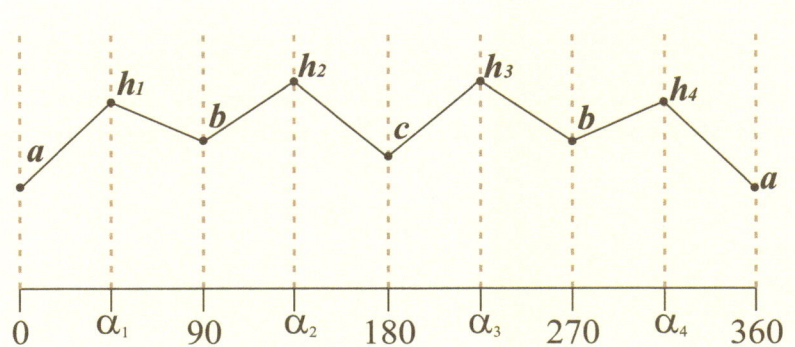

Fig. 16.7 Graphical representation of the instability points

For the development of the following expressions will be assumed a vehicle that viewed front is approximately rectangular and that does not suffer significant deformations during the turns.

With:

g: Acceleration of gravity.

m: Vehicle mass.

U: Work done by the vehicle when turning.

Consider the following cases:

(a) The vehicle rises from the ground, but fails to take a turn of 90° and returns to the starting position:

$$U < mg(h_1 - a)$$

$$U < mg\left(\sqrt{a^2 + b^2} - a\right) \tag{16.9}$$

(b) The vehicle takes a turn of 90°:

$$U = mg(h_1 - a)$$

$$U = mg\left(\sqrt{a^2 + b^2} - a\right) \tag{16.10}$$

(c) The vehicle takes a turn of 180°:

$$U = mg(h_1 - a + h_2 - b)$$

$$U = mg\left(\sqrt{a^2 + b^2} - a + \sqrt{b^2 + c^2} - b\right) \tag{16.11}$$

(d) The vehicle takes a turn of 270°:

$$U = mg(h_1 - a + h_2 - b + h_3 - c)$$
$$U = mg\left(\sqrt{a^2 + b^2} - a + \sqrt{b^2 + c^2} - b + \sqrt{b^2 + c^2} - c\right)$$

$$U = mg\left(\sqrt{a^2 + b^2} + 2\sqrt{b^2 + c^2} - a - b - c\right) \tag{16.12}$$

(e) The vehicle takes a turn of 360°:

$$U = mg(h_1 - a + h_2 - b + h_3 - c + h_4 - b)$$

$$U = mg\left(\sqrt{a^2 + b^2} - a + \sqrt{b^2 + c^2} - b + \sqrt{b^2 + c^2} - c + \sqrt{a^2 + b^2} - b\right)$$

$$U = mg\left(2\sqrt{a^2 + b^2} + 2\sqrt{b^2 + c^2} - a - 2b - c\right) \tag{16.13}$$

Application: Thus, for example, if the vehicle's overturning occurs solely by the action of kinetic energy, the minimum speed at which the turn of 90° occurs, can be calculated as follows, using Eq. 16.10:

$$\frac{1}{2}mv^2 = mg\left(\sqrt{a^2 + b^2} - a\right)$$

$$v = \sqrt{2g\left(\sqrt{a^2 + b^2} - a\right)} \tag{16.14}$$

16.2 "Flip"

When the vehicle rotates around its side axis because its front wheels for some reason are locked, it is said that a movement of pitch or "Flip" (see Fig. 16.8).

To study this type of turn, consider the Fig. 16.9.

Where:

a: distance from the pavement to the center of mass of the vehicle.

c: distance from the roof to the center of mass of the vehicle.

Fig. 16.8 Flip

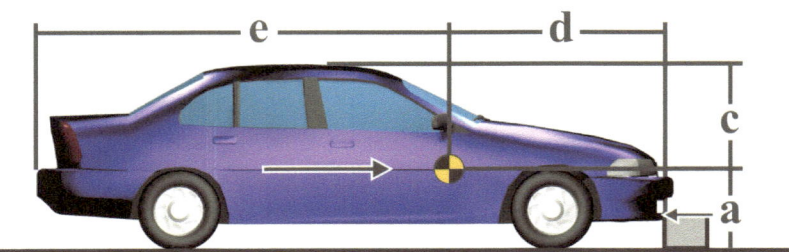

Fig. 16.9 Position of vehicle center of mass

d: distance from the front to the center of mass of the vehicle.

e: distance from the back to the center of mass of the vehicle.

First of all, it is necessary to mention that for the Flip occurs it is necessary that the force blocking the tires is located below the center of mass (Fig. 16.9).

The case of Flip is easier to study than that of the rollover since there are only two points of instability: First, when the vehicle is standing on its front (for which the maximum potential to overcome will be **d** and second, when is standing at its back, for which the maximum potential to overcome will be **e**.

Because most vehicles have the center of mass closest to the front than the back, it is much easier for the vehicle to rotate around its front to rotate around its back as the energy needed for the latter movement is much greater.

16.3 "Spin"

A "Spin" occurs when a vehicle rotates around its vertical axis (Fig. 16.10).

Normally, a car that spin has no rollover or flip because the terrain does not allow it.

It should be remembered that the issue of the kinetic energy of a rotating body has already been discussed in Chap. 5 and very important also remember, that the center of mass during a 360^0-flip movement remain in a straight line.

From Agatha Christie's Poirot, to Reflect and Conclude:

Fig. 16.10 Spin

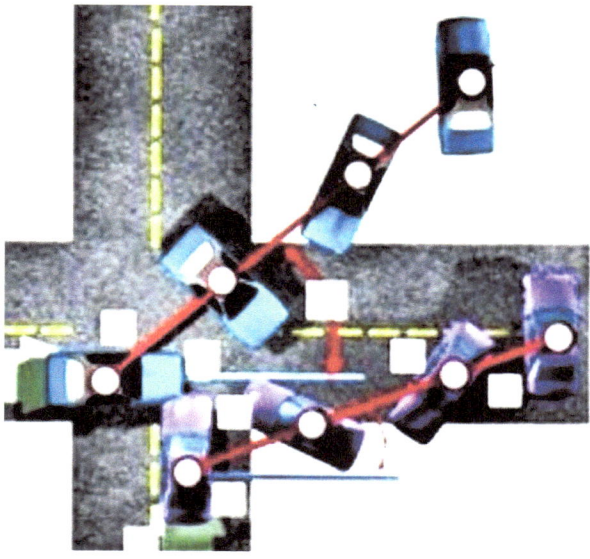

Evil Under The Sun
Its worst problem is that it complicates the simple.

The Adventure of the Italian Nobleman
On the Other Side of the Mirror: Right is Left and Left is Right.
   Remark: See figures 1.1 and 1.2.

Mrs. Macguist is Dead
1. If he has answers for everything, I don't know why he bothers to ask questions.

....

2. I want to know, in spite of all of you.

....

3. You can't avoid the inevitable forever.

Fundamental Equations of the Movement of a Rigid Body

In this chapter we will deduce the translational and rotational motion equations in three dimensions of a rigid body. With these equations it is possible to determine the position of a vehicle in space, as well as its orientation to a global coordinate system at any given instant. In addition, in the case of collisions theory applied to vehicular collisions, some simplifications of this equations are made, as is usual in the automobile practice.

17.1 Euler Equations

The basic equations that describe the three-dimensional motion of a rigid body are, the second Newton law and the equations of angular moment:

$$\sum F = m \cdot a_c \tag{17.1}$$

$$\sum M_C = \dot{H}_c \tag{17.2}$$

where:

$\sum F$ The result of all external forces.
a_c Acceleration of the center of mass.
m Rigid body mass.
$\sum M_c$ The resulting moment relative to the center of mass of all external forces and moments acting on the body.

L. G. Mejía Cañas, *Introduction to the Theory of Vehicular Collisions*, Synthesis Lectures on Mechanical Engineering, https://doi.org/10.1007/978-3-031-62355-4_17

H_c The resulting moment relative to the center of body mass of all linear moments
of the system (the angular momentum of the rigid body relative to its center of
mass).

Next, an expression for the derivative with respect to time of the angular momentum
vector $\dot{H}_c$ will be developed to replace it in Eq. 17.2.
   Let's assume that we have (see Figs. 28 and 38):

(a) A rigid body that rotates with an angular velocity ω and
(b) An inertial frame of reference with coordinates XYZ.
(c) A rotating coordinate frame xyz that rotates with an absolute angular velocity Ω and
   will be fixed to the center of mass of the rigid body, and that seeks to investigate the
   variations of the body's inertial properties over time. The angular velocity Ω may be
   different from the angular velocity ω of the rigid body

The speed of change of H_c measured with respect to the rotatory frame of reference $x\,y\,z$
and the speed of change of the rotating frame with respect to the inertial fixed frame XYZ
for the more general three-dimensional movement are related as follows:

$$\left(\dot{H}_c\right)_{XYZ} = \left(\dot{H}_c\right)_{xyz} + \Omega \times H_c \qquad (17.3)$$

in which:

$\left(\dot{H}_c\right)_{XYZ}$ The derivative with respect to the time of the centroidal angular momentum
of the rigid body, measured from a fixed (absolute) frame of reference.

$\left(\dot{H}_c\right)_{xyz}$ The derivative of the time of the centroidal angular momentum of the rigid
body, measured from a moving (relative) frame of reference.

Ω The angular speed of the mobile (rotatory) frame xyz.

Replacing Eq. 17.3 in Eq. 17.2, we get:

$$\sum M_c = \left(\dot{H}_c\right)_{xyz} + \Omega \times H_c$$

Were

$$\sum M_c = \left(\sum M_c\right)_x \mathbf{i} + \left(\sum M_c\right)_y \mathbf{j} + \left(\sum M_c\right)_z \mathbf{k}$$

$$\Omega = \Omega_x \mathbf{i} + \Omega_y \mathbf{j} + \Omega_z \mathbf{k}$$

$$H_c = (H_c)_x \mathbf{i} + (H_c)_y \mathbf{j} + (H_c)_z \mathbf{k}$$

$$\left(\dot{H}_c\right)_{xyz} = \left(\dot{H}_c\right)_x \mathbf{i} + \left(\dot{H}_c\right)_y \mathbf{j} + \left(\dot{H}_c\right)_z \mathbf{k}$$

Now, considering that the resulting components of the angular momentum vector are:

$$(H_c)_x = I_{xx}\omega_x - I_{xy}\omega_y - I_{xz}\omega_z$$

$$(H_c)_y = -I_{yx}\omega_x + I_{yy}\omega_y - I_{yz}\omega_z$$

$$(H_c)_z = -I_{zx}\omega_x - I_{zy}\omega_y + I_{zz}\omega_z$$

We obtain:

$$\begin{bmatrix} \left(\sum M_c\right)_x \mathbf{i} \\ \left(\sum M_c\right)_y \mathbf{j} \\ \left(\sum M_c\right)_z \mathbf{k} \end{bmatrix} = \begin{bmatrix} \left(I_{xx}\dot{\omega}_x - I_{xy}\dot{\omega}_y - I_{xz}\dot{\omega}_z\right)\mathbf{i} \\ \left(-I_{yx}\dot{\omega}_x + I_{yy}\dot{\omega}_y - I_{yz}\dot{\omega}_z\right)\mathbf{j} \\ \left(-I_{zx}\dot{\omega}_x - I_{zy}\dot{\omega}_y + I_{zz}\dot{\omega}_z\right)\mathbf{k} \end{bmatrix} + \begin{vmatrix} \mathbf{i} & \mathbf{j} & \mathbf{k} \\ \Omega_x & \Omega_y & \Omega_z \\ (H_c)_x & (H_c)_y & (H_c)_z \end{vmatrix} \qquad (17.4)$$

However, if the x y z axes are chosen to match the principal axes of inertia of the rigid body, then the products of inertia are equal to zero:

$$I_{xy} = I_{yx} = 0 \, I_{xz} = I_{zx} = 0 \, I_{yz} = I_{zy} = 0 \qquad (17.5)$$

and therefore, the Eq. 17.4 can be simplified as follows:

$$\begin{bmatrix} \left(\sum M_c\right)_x \mathbf{i} \\ \left(\sum M_c\right)_y \mathbf{j} \\ \left(\sum M_c\right)_z \mathbf{k} \end{bmatrix} = \begin{bmatrix} (I_{xx}\dot{\omega}_x)\mathbf{i} \\ (I_{yy}\dot{\omega}_y)\mathbf{j} \\ (I_{zz}\dot{\omega}_z)\mathbf{k} \end{bmatrix} + \begin{vmatrix} \mathbf{i} & \mathbf{j} & \mathbf{k} \\ \Omega_x & \Omega_y & \Omega_z \\ (H_c)_x & (H_c)_y & (H_c)_z \end{vmatrix} \qquad (17.6)$$

And writing this vector relationship in terms of its scalar components:

$$\left(\sum M_c\right)_x = I_{xx}\dot{\omega}_x - I_{yy}\Omega_z\omega_y + I_{zz}\Omega_y\omega_z$$

$$\left(\sum M_c\right)_y = I_{yy}\dot{\omega}_y - I_{zz}\Omega_x\omega_z + I_{xx}\Omega_z\omega_x$$

$$\left(\sum M_c\right)_z = I_{zz}\dot{\omega}_z - I_{xx}\Omega_y\omega_x + I_{yy}\Omega_x\omega_y$$

Now, in order to get further simplification, a reference frame xyz will be chosen rotating at the same angular velocity as the rigid body, that is $\Omega = \omega$. Then:

$$\left(\sum M_c\right)_x = I_{xx}\dot{\omega}_x - I_{yy}\omega_z\omega_y + I_{zz}\omega_y\omega_z$$

$$\left(\sum M_c\right)_y = I_{yy}\dot{\omega}_y - I_{zz}\omega_x\omega_z + I_{xx}\omega_z\omega_x$$

$$\left(\sum M_c\right)_z = I_{zz}\dot{\omega}_z - I_{xx}\omega_y\omega_x + I_{yy}\omega_x\omega_y$$

Finally, by grouping terms and considering the scalar components of Eq. 17.1, the so-called Euler equations are obtained:

Euler Equations of motion (17.7)

$\left(\sum M_c\right)_x = I_{xx}\dot{\omega}_x - \left(I_{yy} - I_{zz}\right)\omega_y\omega_z$	$\sum F_x = m(a_c)_x$
$\left(\sum M_c\right)_y = I_{yy}\dot{\omega}_y - (I_{zz} - I_{xx})\omega_z\omega_x$	$\sum F_y = m(a_c)_y$
$\left(\sum M_c\right)_z = I_{zz}\dot{\omega}_z - \left(I_{xx} - I_{yy}\right)\omega_x\omega_y$	$\sum F_z = m(a_c)_z$

Historical Note: The six equations above are known as the Euler's equation of motion, in honor of the 18th-century Swiss mathematician Leonhard Euler (1707-1783).

17.2   Euler Angles

So far, with the first three rotation equations of Euler it is possible to find the components of the angular velocity of the body ω_x, ω_y and ω_z with respect to the axes xyz (main axes of inertia of the rigid body, see Eq. 17.5). Now, in order to determine the orientation of the body, three more equations are defined that express the angular velocities ω_x, ω_y and ω_z in terms of Euler's angles, as explained below. The coordinate systems that accompany Figs. 17.2, 17.3, 17.4, 17.5 and 17.6 prepared by the author are very helpful and correspond to the methodology indicated in reference [21].

- Initially, let's choose a reference (initial) position in which the axes xyz and XYZ match (Fig. 17.1).
- The axes xyz are then rotated an angle ψ (precession angle) on the axis Z, z as shown in Fig. 17.2.

We will call the new resulting axis x´ y´ z´. Therefore, the vector ˙ψ represents the reason of change of the precession angle ψ with respect to time it has the same direction as the axis z´, which is the same axis Z.

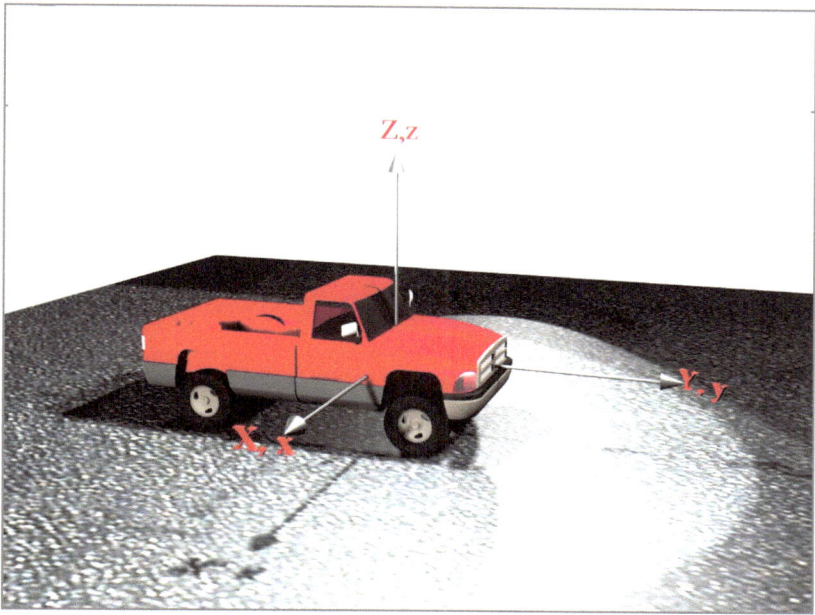

Fig. 17.1 Reference position

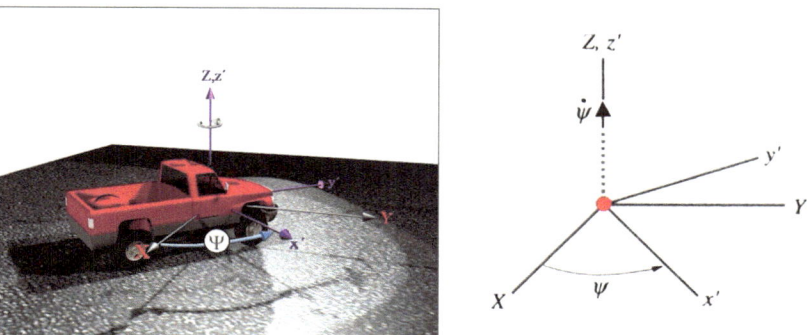

Fig. 17.2 Rotation about axis Z, z

- The next step is to rotate the x' y' z' axes an angle θ (nutation angle) on the x' axis generating the x" y" z" axes as shown in Fig. 17.3.

In this way, the vector $\dot{\theta}$ represent the reason for changing the nutation angle θ with respect to time it has the same direction as the axis x" (which is the same axis x').It

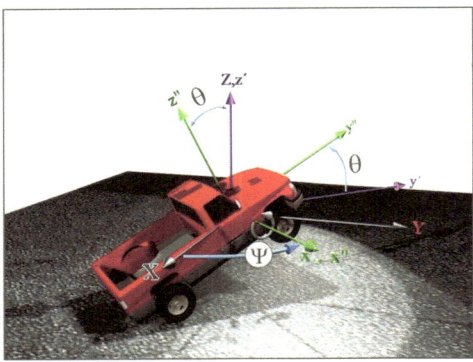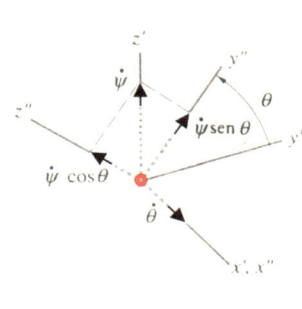

Fig. 17.3 Rotation about axis x

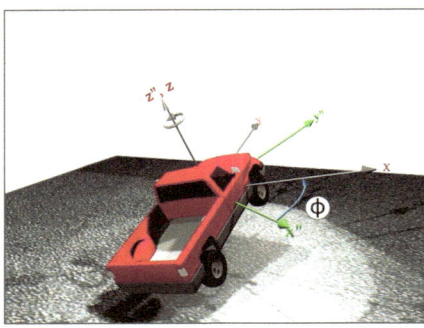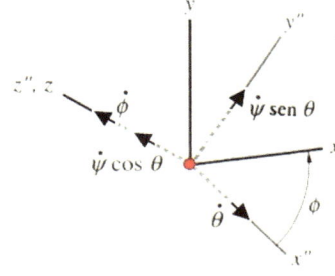

Fig. 17.4 Rotation about axis z"

Fig. 17.5 Decomposition of
$\dot{\psi}\sin\theta$ on the axes x, y

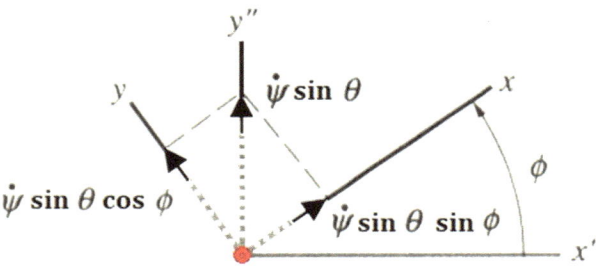

is easy to see that the components of vector $\dot{\psi}$ in axis z" and y" are $\dot{\psi}\cos\theta$ and $\dot{\psi}\sin\theta$ respectively.

Finally, the axes x" y" z" are rotated an angle ϕ (angle of rotation) on the axis z" (see Fig. 17.4) finally getting the axes x y z (main axes of body inertia). So, the vector $\dot{\phi}$

representing the reason for changing the angle of rotation ϕ regarding time has the same direction that the axis z" (which is the same axis z).

From the right Fig. 17.4, it can be seen that only the vectors $\dot{\psi}\sin\theta$ and $\dot{\phi}$ are to be resolved into their components on the main axes of inertia (x y z). However, both vectors are on the plane x" y" which is the same plane x y, then there will be no component of these on the axis z.

The Fig. 17.5 shows the vector $\dot{\psi}\sin\theta$ about the plane x y (the z axis is pointing out of the page) and is easy to see that component on the x axis is $\dot{\psi}\sin\theta\sin\phi$ and the component of the axis y is $\dot{\psi}\sin\theta\cos\phi$.

The Fig. 17.6 shows the vector $\dot{\theta}$ about the plane x y (the axis z is pointing out the page) and easily see that component on the axis x es $\dot{\theta}\cos\phi$ and that the component on the axis y is $\dot{\theta}$.

And the Fig. 17.7 in turn shows all the components of angular velocity in the three coordinate directions determined by Euler angles:

Fig. 17.6 Decomposition of $\dot{\theta}$ on the axis x, y

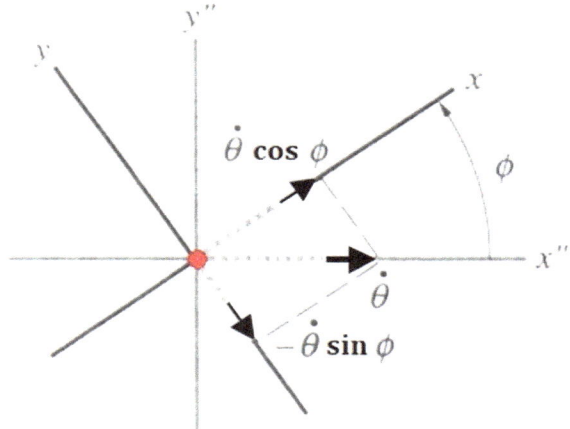

Fig. 17.7 Components of angular velocity in three coordinates directions in function of Euler angles

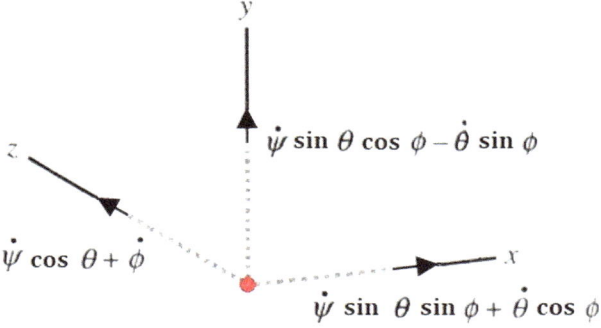

Then, the three equations that express ω_x, ω_y and ω_z in terms of Euler's angles are:

$$\omega_x = \dot{\psi}\sin\theta\sin\phi + \dot{\theta}\cos\phi$$

$$\omega_y = \dot{\psi}\sin\theta\cos\phi - \dot{\theta}\sin\phi \qquad (17.8)$$

$$\omega_z = \dot{\psi}\cos\theta + \dot{\phi}$$

So, like the terms, ω_x, ω_y and ω_z are known, there are three equations in which the three unknowns are the Euler angles: θ, ψ, ϕ which can be determined and so, it is possible to know the orientation of the body, whether it's a car or a spaceship.

Accordingly [17], in aerodynamics other orientation angles are used to avoid numerical difficulties that can appear when θ is zero. Such angles are called: "Roll", "Pitch" and "Yaw" (Fig. 17.8).

An aircraft can modify its position, but the mathematical transformation of the inertial frame to the body frame requires that all three rotations be done in the following order: First an angle ψ_1 (Yaw) is rotated around the vertical axis, then an angle ψ_2 (pitch) should be rotated around the new wing axis and finally a roll an angle ψ_3 around the longitudinal axis should be made.

Then, performing a procedure similar to that carried out with the Euler angles, three equations are obtained that replace the three equations previously inferred.

$$\omega_x = \dot{\psi}_3 - \dot{\psi}_1\sin\psi_2$$

$$\omega_y = \dot{\psi}_2\cos\psi_3 + \dot{\psi}_1\cos\psi_2\sin\psi_3 \qquad (17.9)$$

$$\omega_z = -\dot{\psi}_2\sin\psi_3 + \dot{\psi}_1\cos\psi_2\cos\psi_3$$

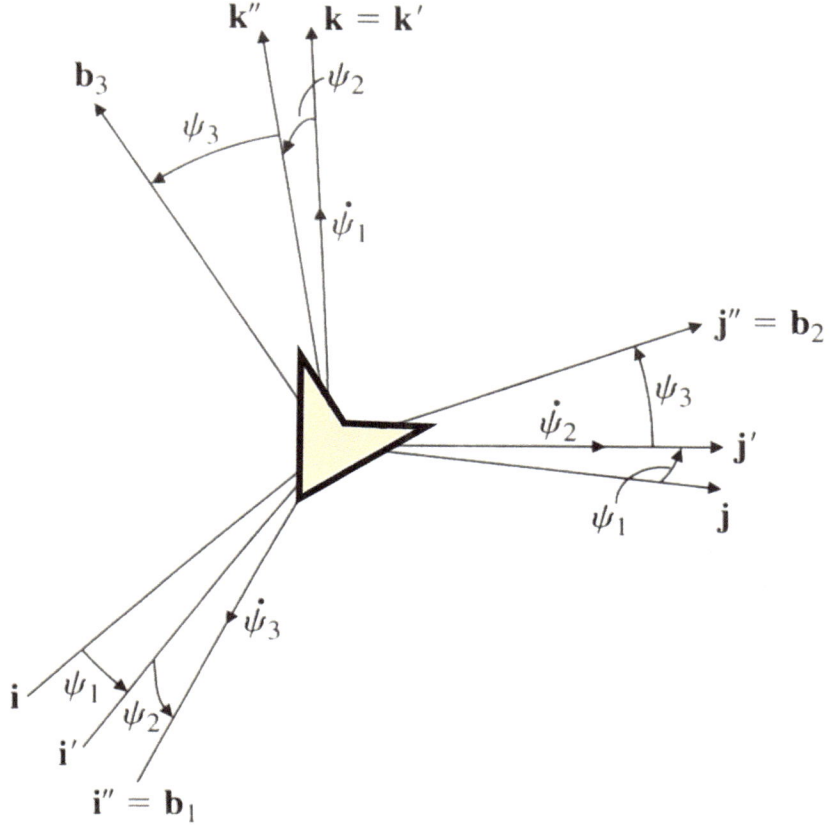

Fig. 17.8 Orientation angles used in aerodynamics

17.3   Simplification of Euler Equations for Automobiles

In general, the axis for the analysis of vehicular collisions are the axis shown in Fig. 17.9. Considering this axis system, it is well known, that the inertia moments Ix-x and Iz-z are roughly the same. This is because, for most vehicles, the mass distribution from the rotation axes is similar in both cases.

At the same time, the moment of inertia on the axle y-y is much smaller than the other two moments of inertia on the main axes, for this reason, there is greater rotation when a torque is applied to the term Iy-y that when applied to Ix-x or Iz-z.

For vehicles it is possible to simplify Euler's equations found in numeral 17.1. Considering that:

- $I_{xx} \approx I_{zz}, \Rightarrow I_{xx} - I_{zz} \approx 0$

Fig. 17.9 Axis system

- $I_{yy} \ll I_{xx}, \Rightarrow I_{xx} - I_{yy} \approx I_{xx}$
- $I_{yy} \ll I_{zz}, \Rightarrow I_{zz} - I_{yy} \approx I_{zz}$

Then, by applying these findings, the Eq. 17.7 become into Eq. 17.10:

Simplified Euler Equations of motion (17.10)	
$\left(\sum M_c\right)_x = I_{xx}\dot{\omega}_x + I_{zz}\omega_y\omega_z$	$\sum F_x = m(a_c)_x$
$\left(\sum M_c\right)_y = I_{yy}\dot{\omega}_y$	$\sum F_y = m(a_c)_y$
$\left(\sum M_c\right)_z = I_{zz}\dot{\omega}_z - I_{xx}\omega_x\omega_y$	$\sum F_z = m(a_c)_z$

Rocket Equation

To consider mass losses present in many collisions (parts of the vehicle fly off), it is necessary to use the well-known "Rocket equation", developed by the Russian physics Konstantin Tsiolkovsky in 1897, based in the concept of conservation of momentum, which expresses that any vehicle, for example a rocket, can accelerate himself ejecting mass a high velocity in opposite sense. Then, at the end, an analogy is made between the two situations: the vehicle collision and the rocket movement.

Suppose that in the instant t we have a rocket of total mass m moving with speed v, regarding the earth and besides that in a time interval dt the rocket expels a mass of gases dm_g with speed v_s regarding the rocket itself (and, therefore, with speed v-v_s regarding the land). Let dv y dm the changes in speed and mass that the rocket undergoes as a consequence of the mass of gases ejected mass.

Accordingly of the principle of conservation of momentum, the force applied to a body equals the rate of change of its momentum. The total momentum before expelling the gases is mv and the total moment after expelling them is the rocket's momentum $(m + dm)(v + dv)$, plus the momentum of the expelled gases that is given by the expression $dm_g (v$-$v_s)$, where the mass of gases is denoted by dm_g.

With the above considerations it is possible to obtain the so- called "Rocket equation":

$$mdv + v_s dm = 0 \tag{18.1}$$

$$dv = -\frac{v_s}{m}dm$$

L. G. Mejía Cañas, *Introduction to the Theory of Vehicular Collisions*, Synthesis Lectures on Mechanical Engineering, https://doi.org/10.1007/978-3-031-62355-4_18

Now, integrating during the interval in which there is mass loss, the desired speed is reached:

$$v = -v_s \ln\left(\frac{m}{m_0}\right) \qquad (18.2)$$

where m_0 is the total initial mass of the rocket, namely the mass of the rocket plus that of the fuel, at the moment it takes off.

This equation can be used similarly in a vehicular collision, where m_0 would be the mass of a vehicle before the crash, m mass after collision, v_s the speed at which the lost mass is ejected and so v would be the change in the speed of the vehicle, due to the loss of mass. The minus sign indicates that this change in speed is presented in the opposite direction to the losing mass.

With this very short discussion, it is possible to see the great complexity that arises if necessary to consider all the factors present in a vehicle collision, including the variable mass due to loss of parts of the vehicle caused by the collision itself.

It should be noted that there are many other factors that may be important when studying vehicle collisions, however, with the state of current knowledge and despite the help of computers, it is almost impossible to take them all into account. For example, the phenomenon of the flutter, present in the front wheels of the vehicles, which requires additional considerations and are mentioned only to emphasize the complexity of attempting to conduct an "exact" collision investigation. To consider the flutter in a crash investigation, following aspects should be considered:

(a) A tilting in a vertical plane that causes an inclination of the axle in elevation, caused by the vertical elasticity of the tires denoted by angle ϕ (angle of tramp).
(b) A sidewise shifting of the axis is also indicated at a magnitude x, due to the lateral flexibility of shock absorbers.
(c) And looking from top, a flutter of the wheels can appear referred as shimmy, which is denoted by angle ψ.

Then we have three degrees of freedom: ϕ, x, ψ which are coupled, that is if either of them is present, so are the other two.

The mathematical analysis of these three coupled movements, in the most elementary of cases, making as many simplifications as possible, leads to the use of sixth-degree frequency equations and eighth-degree equations without making simplifications. Therefore, the study of this phenomenon is not simple, and if required by a specific situation, it would greatly complicate the study of any collision, but fortunately, it is rarely necessary.

© The Author(s), under exclusive license to Springer Nature Switzerland AG 2025   101
L. G. Mejía Cañas, *Introduction to the Theory of Vehicular Collisions*, Synthesis Lectures on Mechanical Engineering, https://doi.org/10.1007/978-3-031-62355-4_19

Ethical and Other Considerations for Small Vehicles (SV)

With the growth of cities, the number of *Small Vehicles* (**SV**) such as motorcycles, bicycles, scooters, skateboards etc. has been increasing, as they not only are a very good solution to mobility but also contribute to solve the pollution problems caused by *Big Vehicles* (**BV**) such as automobiles, buses and trucks.

Like any other vehicle, the SV can be easily involved in collisions with other vehicles or pedestrians, so it is important to make some comments that aim to reduce these situations, in order to achieve, that its intelligent and safe use, serve for benefit of cities and their inhabitants.

20.1 Introduction

Since ancient times, the humanity sought to understand nature from not only the regularities that occurred in the phenomena that influenced their lives, such as the determination of the winter and summer seasons to sow and collect the food that gave them life, but to understand the movement of the stars, which, for them, as still today, represented a fascinating spectacle.

This search for answers has been called natural philosophy and led to the first steps in the field of mathematics, astronomy and what today we call science, that is, the study of the laws governing the phenomena of nature.

After this time full of intellectual effervescence, during the middle age more than 1000 years of apparent stillness came, period in which, fortunately, the Arab erudite and the monks in the abbeys saved the whole treasure of ancient knowledge. In the renaissance,

103
L. G. Mejía Cañas, *Introduction to the Theory of Vehicular Collisions*, Synthesis Lectures on Mechanical Engineering, https://doi.org/10.1007/978-3-031-62355-4_20

in this rebirth of the sciences and the arts, the mathematical formulation of these natural phenomena was initiated, which allowed them to be defined clearly and accurately, moving them away from mystical interpretations, achieving exceptional advancement in physics and especially of the branch that studies the laws of movement and equilibrium: the mechanics, with which the way was opened for the mathematical formulation of collision theory.

20.2   But is This Story of Some Interest for SV Drivers?

The answer is yes. As already mentioned, in general, physical phenomena can be expressed by mathematical equations, which are met in all cases, without exception, and those who violate them can become crippled, dead or cause serious harm to others.

To prevent the above from happening, it isa necessary above all, to have a "driving culture" that allows SV drivers to use their means of transport safely for their good and that of others, avoiding reaching risky situations.

This "driving culture" should start from the minimum understanding of basic physical laws, which we will try to explain in a simple way below, considering three aspects, which, for their own good, SV drivers should know:

(a) The concept of the stop distance,
(b) The enormous forces that occur during a collision and
(c) The factors that ease the occurrence of accidents and aggravate their consequences.

20.3   Stop Distance (Braking Distance)

The first thing to understand is that the action of braking, if a situation that requires it to avoid a collision with a vehicle, an object or with a pedestrian, depends on two factors:

- The psychotechnical time and
- The **braking time** itself.

At the same time, the psychotechnical time, also known as "perception-reaction" time is the sum of two times:

(a) *The perception time*, which varies between 0.5 s and 2.0 s, is the time it takes the brain to *understand* the situation to *decide* about what is the best way to follow to avoid the unforeseen situation and to *order* the body, for example, whether it should brake or not or if it should better turn, etc. and,

(b) *The reaction time*, which varies between 0.5 s and 1.0 s, is the time it takes the driver to execute the action ordered by the brain.

In general, it is considered that the minimum psychotechnical time is 1.5 seconds for young people, rested and without influence of alcohol or drugs, being important to note, that the use of cell phones while driving, can increase this time by 1.0 second which can lead to a catastrophe. In relation to this point, one of the first tasks that the person who is investigating a vehicle accident must perform is to check the cell phone or cell phones of those involved, to determine whether they were using it at the time of the accident, because, usually, there is a high probability that this fact will be the cause and trigger event of the accident.

This time can also increase if the rider has a helmet hanging on his arm, as this makes the braking maneuver difficult. It is important to note also, that the braking time is not instantaneous, but responds to a physical law that depends on the speed of the SV and the friction that occurs between the tires and the track.

In this way, the stop distance that the SV travel on a level track from the time the obstacle (pedestrian, parked vehicle, construction on the track, animal crossing the track) is seen and the vehicle stops completely, is given by the following expression, in which the first term corresponds to psychotechnical time and the second to the braking time:

$$\text{Stop distance} = vt + v^2/2gf \qquad (20.1)$$

Let's clarify this formula with an example: if a motorcyclist is driving at a speed "v" from 60.0 kilometers per hour (approximately 37.3 mil/h), or what is the same at 16.7 meters per second (approximately 54.8 ft/s), what it means that in a second the SV travel 16.7 meters(54.8 ft/s), also in a second the vehicle travel 16,7m (54.8ft), a huge distance, on a level, paved, dry track, for which the friction factor "f" determined by trials is equal to 0.8 and psychotechnical time "t" is 1.5 seconds, the stop distance, that is, the distance required to brake completely is 43.0 meters, which means that if the distance between this motorcycle and the vehicle ahead, the one that for some reason suddenly has braked, is less than 43.0 meters, without escape, the motorcycle will violently collide with the vehicle standing still.

If the driver of the motorcycle, in these same conditions is using a cell phone, the safety distance that should be kept to avoid the collision with the obstacle, would be minimum 60.0m.

Next is a table that allows find the "stop distances", for various speeds, considering a psychotechnical time of 1.5 seconds, a paved track, flat and dry with a friction factor of 0.8, without any obstacle that impairs visibility, by day and without the driver of the bike or motorbike, go using the cell phone:

Speed (km/h)	Stop distance (m) (Breaking distance)
5	2
10	5
20	10
30	17
40	25
50	33
60	43
80	65
100	91

The "braking distance" values in the table above may be more than double those showed in this table in case the pavement is wet or there is little visibility, for example, at night or on track slope and with little lighting. If these conditions are combined with the use of a cell phone while driving, the consequences will undoubtedly be catastrophic. Equal situations of increased risk, in relation to normal driving conditions, occur when driving at the same time or after using hallucinogens.

20.4   "Violence" of a Collision

The understanding of what was happening in a collision was achieved about 300 years ago, thanks to the English physicist Isaac Newton. Collisions can be classified in different ways and it analysis can present several degrees of difficulty, however, no matter what the collision is, common patterns are always found.

From our daily experience we know that a moving body can cause damage, which it would not cause if it were at rest. This property of moving bodies, of causing damage is what is known as "Kinetic Energy" which is expressed in a unit called Jules.

In physics, kinetic energy is defined by the following mathematical expression, in which "m" is the mass of the vehicle and its passengers and "v" its speed:

$$E_c = \frac{1}{2}mv^2 \tag{20.2}$$

Two very important teachings are obtained from this expression:

(a) Doubling the speed quadruples the kinetic energy, i.e. quadruples a vehicle's ability to cause damage, no matter if it is a bicycle or a truck and

(b) The higher the combined vehicle-passenger weight, the greater the adverse consequences that occur during a collision, and it should be clear that the damage ability
is greater if the SV carries a passenger or 2 or more passengers or transports goods.

Sometimes the destructive power, the ability to do harm that entails the concept of
"Kinetic Energy" expressed in Jules, is difficult to understand, but to clarify it we can
make use of an equivalence that exists between this kinetic energy and another form of
energy called "Potential Energy", which is expressed by the following equation, in which
"m" es la masa, "g" the acceleration of gravity and "h" is the height at which the mass
"m" is located:

$$E_p = mgh \tag{20.3}$$

By matching these two forms of energy, an equivalence can be found between the
kinetic energy (violence) present in a collision and that which is reached when a person
falls from a building:

| Motorcycle speed in km/h | 30 | 40 | 50 | 60 | 80 | 100 |
(mi/h)	18.6	24.8	31.0	37.3	49.7	62.1
Number of equivalent fall floors	1	2	3	5	8	13

What this table indicates, is that the impact that a motorcyclist receives in a collision,
for example traveling at 60 kilometers per hour, would be equivalent to if the motorcyclist
decided to launch from a fifth floor of a building, and if he travel at 100 km/h, it is as
if he decides to launch from a 13th floor, and, for obvious reasons, it is very likely that
the first one is left dead or disabled and the second, no doubt, will be dead .Those who
roar their motorcycles, especially at night, would well do that for themselves and their
families, they would have in mind the above information.

20.5   Civil and Penal Responsabilities

Although it is not the primary objective of a book on vehicle collisions to address issues
related to civil and criminal responsibilities that arise in vehicular collisions, the acceler-
ated growth of these means of transport, coupled with the growing lack of culture when
driving of those who use them, especially motorcycles, has led to their risky use, for
drivers and pedestrians who share with them the tracks and walkways.

From a point of view of physical law, reckless driving SV's will drivers of these
vehicles who engage in collisions, in which other vehicles or pedestrians have no time to
react (see perception-reaction time) civilly and criminally liable.

Three common cases are mentioned in this regard:

1. As mentioned in connection with the psychotechnical time, the zigzagging on the track changing lanes, is a maneuver that takes less than a second and any driver of a motor vehicle that is driving correctly, does NOT have time to perform the braking maneuvers to avoid the collision (see numeral 20.3) and the motorcyclist or cyclist, can be hit by the motor vehicle, without any responsibility on the part of the driver of that vehicle from a point of view of the aforementioned laws of physics (see Eq. 20.1).
2. The same is true, when a cyclist or motorcyclist does not stop at an intersection, continues the march and turns, assuming that, because of the small size of the bike, the vehicle that comes with the track will have time to brake so as not to hit it. Well, it's not like that, and the motorcyclist or cyclist is making a serious mistake, which can lead to being run over, without the driver running him over having any responsibility in this accident.
3. Another typical situation of risk of a serious accident occurs on non-dual carriageway roads, in which motorcyclists or cyclists who are going at high speed get out of the lane, invading, even a little, the lane of vehicles traveling on the other direction.

20.6   Circumstances that Facilitate the Occurrence of Accidents and Amplify Their Consequences

Several factors can facilitate the occurrence of vehicular accidents and may also aggravate their consequences. Although some have already been mentioned, it is worth to repeat them:

- Using alcohol or hallucinogens
- The condition of the road
- Driving at night
- Using your cell phone while driving
- Charging the helmet hanging in one hand
- Carrying a tender puppy in your arms
- Overload the SV, carrying, for example, two passengers in addition to the driver
- Zigzagging on the track
- Do not stop at intersections when turning
- Taking curves with excessive speed
- And first of all, the lack of a driving culture.

Because of their importance, at this point it is worth repeating a paragraph of Chap. 6, which draws attention that SV's although have very little mass, for its speed can also cause serious accidents by injuring or causing the death of pedestrians who drive unsuspectingly on the streets or are crossing the tracks, as these, they are the weakest and most exposed part of a city's entire mobility system. *This point is vitally important for those SV drivers, who often drive at high speeds in pedestrian areas.*

20.7   The City, Bicycles and the Motorcycles

The use of SV's is of enormous importance for modern cities, characterized by their high population density, congestion and pollution. The success of the use of these vehicle systems does not occur on its own, but is the result of the sensible participation of those who use them and the municipal administrations.

The participation of the SV's drivers in making these transport systems work properly is mainly limited to compliance with established laws and driving with sensibly and culture, but nothing is going to be achieved, if the municipal administrations do not act quickly, taking steps so that these means of transport, instead of being a solution for the city, become a dangerous new headache.

Among other things, it is up to the city to plan the circulation of these vehicles by taking the following minimum measures (photos of the author):

1. Building cycle paths with low separators (Photo 20.1) and not with so-called milestones, used in some cities, as these distract drivers and cause accidents. (Fig. 20.1)
2. In multi-lane riders, building special lanes for SV's, obviously with a smaller width than standard vehicle lanes (Photos 20.1 and 20.2).
3. Properly signaling each of the lanes to avoid conflicts and accidents and clearly showing who has the priority in each case (Photos 20.1 and 20.2), not forgetting that pedestrians always have preference.
4. On the most demanded roads, demarcating an area for motorbikes of about 10 m in length, from the traffic lights, so that when the traffic light changes, they can continue their march in a more fluid way, which in turn will have an impact on a more orderly and faster circulation of the other vehicles.
5. Planning with due anticipation the construction of large SV's parking areas, because if the use of these vehicles grows, as is desirable, the parking sites will require very important areas (Photos 20.3 and 20.4).
6. Teaching cyclists their duties and rights (Photo 20.5).
7. Trying not to strangle the vehicular lanes, planning the SV's lanes, for example, through the tree-covered areas where the water streams flow (Photo 20.6).
8. Taking measures to allow SV's to be transported on public transport (see Photos 20.7 and 20.8).
9. Regulations that the SV's, especially motorcycles, bicycles and scooters to be used in the city are factory equipped with a speed reducer, which does not allow them to circulate more than, for example, 50 km/h.
10. Imposing severe fines to lawbreakers.

Fig. 20.1 Bikes for everyone

Photo 20.1 Properly demarcated bike with low separators, which do not cause visual contamination (Paris)

Photo 20.2 Ideal coexistence between tram tracks, cars, motorbikes, bicycles and pedestrians (Vienna)

Photo 20.3 Bicycle parking (Biciparks) in Maastricht, Netherlands

Photo 20.4 One-and two-story bicycle parking in Munich

Photo 20.5 Instructor teaching cyclists the traffic rules (Berlin)

Photo 20.6 Author's proposal, for a bike-lane along an open channel in the city of Medellín, Colombia

Photo 20.7 Bicycle transport on the Berlin Metro

Photo 20.8   Unicycle. Paris Metro

In this Chapter: Final Reflection

Laws need to be tightened, given the lack of civility of the majority of those who drive small vehicles (SV)?

In some times it was necessary to do so as is demonstrated by the following two laws of Hammurabi.

If a builder builds a house for a man and does not make its construction firm and the house collapses and causes the death of the owner of the house—that builder shall be put to death. If it destroys property, he shall restore whatever it destroyed, and because he did not make the house firm he shall rebuild the house which collapsed at his own expense. If a builder builds a house for a man and does not make its construction meet the requirements and a wall falls—that builder shall strengthen the wall at his own expense.

—The Code of Hammurabi, c. 2250 B.C.

Taken and adapted from reference [29]

Bibliography

1. Noon Randall K., "Engineering Analysis of Vehicular Accidents"; Ed. CRC Press, USA 1994.
2. Szabó István "Geschichte der Mechanischen Prinzipien"; Ed. Birkhäuser Verlag, 1987.
3. Mach Ernst, "Die Mechanik in Ihrer Entwickelung" F.A. Brockhaus, Leipzig, 1908.
4. Dugas Rene, "A History of Mechanics" Dover Publication Inc. 1988.
5. Gamow, George "Biografía de la Física" Salvat Editores 1971.
6. Lozano Juan Manuel, "Como Acercarse a la Física". Editorial. Limusa; México 2001.
7. Hugh D. Young, "University Physics"; Eight edition, volume 1; Addison Wesley Publishing Company; Inc. 1992.
8. Van Kirk, Donald J. "Vehicular Accident Investigation and Reconstruction". USA, C.R.C. Press. 2001.
9. Winkler, Johannes; Aurich, Horst. "Technische Mechanik", VEB Fachbuchverlag Leipzig.1974.
10. Göldner, Hans; Holzweissig, Franz. "Leitfaden der Technischen Mechanik", VEB Fachbuchverlag Leipzig 1976.
11. Russell, C. Gregory. "Equations & Formulas for the Traffic Accident Investigator and Reconstructionist". Lawyers & Judges Publishing Company, Inc. USA. 1999.
12. Casteel, David A; Moss Steven D. "Basic Collision Analysis and Scene Documentation". Lawyers & Judges Publishing Company, Inc. USA. 1999.
13. (ISSI) Information Systems and Services, Inc. NHTSA "Test Reference Guide" Version 5; Volume 1: Vehicles Tests. Prepared for: U.S. Department of Transportation (DOT); National Highway Traffic Safety Administration (NHTSA) Mayo 2001.
14. Crede, Charles "Conceptos Sobre Choque y Vibración en el Diseño de Ingeniería". Herrero Hermanos Sucs S.A, México D.F 1970.
15. Arya Suresh, O´Neil Michael, Pincus George; "Design of Structures and Foundation for Vibrating Machines" Gulf Publishing Company, Houston Texas 1979.
16. Thomson, William Tyrrel; "Introduction to Space Dynamics", Dover Publications, Inc., New York 1986.
17. Wiesel, William E; "Space Flights Dynamics"; Second Edition; McGraw Hill International Companies, Inc. 1997.
18. Bándres; J.C.; "Seguridad del Automóvil"; Editorial Albatros, Buenos Aires (Argentina.), 1975.
19. López Muñiz, Miguel, "La Determinación de la Velocidad de los Vehículos"; Editorial Gesta; Madrid.
20. Den Hartog, J.P. "Mechanical Vibrations"; Dover Publications, Inc. New York.
21. Bedford - Fowler, "Dinámica. Mecánica Para Ingeniería", Editorial Prentice Hall; México 200.
22. Herrera, Miguel Angel, "Biofísica, Geofísica, Astrofísica. Para qué Sirve la Física", Universidad Nacional Autónoma de México, Fondo de Cultura Económica; México 2001.

© The Editor(s) (if applicable) and The Author(s), under exclusive license
to Springer Nature Switzerland AG 2025
L. G. Mejía Cañas, *Introduction to the Theory of Vehicular Collisions*, Synthesis Lectures
on Mechanical Engineering, https://doi.org/10.1007/978-3-031-62355-4

23. Tomson, William; Dahleg Marie Dilon, "Theory of Vibration with Applications", Prentice Hall; Fifth Edition, USA 1998.
24. Du Puy Thierry, Dumaitre Pierre "Técnicas de utilización de la Moto" Cepadues Editions, Toulouse, Francia, 1999.
25. C. Gregory Russell "Equation & Formulas for the Traffic Accident Investigator and Reconstructionist" Lawyers & Judges Publishing Company, Inc. USA 1999.
26. Das, Braja M.; Kassimali, Aslam; Sami, Sedat. "Mecánica para ingenieros: Dinámica", Southern illinois University at Carbondale. Norienga Editores; México, 1999.
27. Stiglat, Klaus, "Apokalypsebau", 128 pages,2010. Copyright Ernst & Sohn GmbH. Reproduced with permission.
28. Cross Hardy, "ingenieros y las torres de marfil". McGraw-Hill de México, 1971.
29. Gere James, Shah Haresh," TERRA NON FIRMA", WH Freeman and Company. New York. 1984.
30. Knothe, Klaus; Stichel, Sebastian. Schienenfahrzeugdynamik, Springer Verlag, Heidelberg, New York, 2003.